Advance Praise for *Earth-Sheltered Houses*

If there were a Pulitzer prize for building books, Rob Roy would surely have won it by now. His books are always a pleasure to read — witty, entertaining, and informative. This latest volume provides crucial and hard-to-find details about a sticky subject — how to design, build, and waterproof the parts of a house that are in direct contact with the earth. For lack of some of the information in this book, one of my first living roofs has sprung multiple recurring leaks. I'm about to tear it off and replace it the right way — using *Earth-Sheltered Houses* as my guide.

— Michael G. Smith, co-editor of *The Art of Natural Building*
and co-author of *The Hand-Sculpted House*

Rob Roy is the guru the of the earth sheltered home movement. *Earth-Sheltered Houses* is based on years of practical experience building and living in earth sheltered homes. The diagrams, photos, and charts make it a must have resource for anyone considering building with this medium. And better yet, the humor and wisdom of the writing make it a fun read.

— Tehri Parker, Executive Director, Midwest Renewable Energy Association

The voice of vast experience, cordwood guru Rob Roy's new book, *Earth-Sheltered Houses: How to Build an Affordable Underground Home* will be the hands-on bible of earth-coupled construction for the next generation of eco-builders.

— Catherine Wanek, author of *The New Strawbale Home;*
and co-author of *The Art of Natural Building*

Finally, a detailed yet user-friendly book on earth sheltered houses. As always, cordwood-masonry master Rob Roy offers a hands-on, accessible read on the advantages (comfort and savings from thermal mass, living roofs, affordability and dependability) and challenges (insulation and waterproofing) of one of man's earliest and simplest forms of housing.

— Andre Fauteux, publisher/editor, *La Maison du 21e siecle* magazine

At last a comprehensive, authoritative book on earth-sheltered home building. This valuable book contains a wealth of accurate and clearly presented information on one of the most important ways to build environmentally friendly, energy-efficient homes.

— Dan Chiras, author of *The Solar House; The Natural House; The New Ecological Home; and The Homeowner's Guide to Renewable Energy*

If it's "conventional" advice you are seeking — on home design, construction, or financing — Rob Roy is not your man. But if you are receptive to new solutions, Rob Roy is a fountain of new ideas. He is one of the truly original thinkers of our time.

— Stephen Morris, founder of The Public Press and publisher, *Green Living Journal*

Earth-Sheltered Houses

HOW TO BUILD AN AFFORDABLE UNDERGROUND HOME

ROB ROY

NEW SOCIETY PUBLISHERS

Cataloging in Publication Data:

A catalog record for this publication is available from the National Library of Canada.

Cover design by Diane McIntosh. Cover image: Rob Roy.

All photos and original illustrations by Rob Roy, unless otherwise noted.

Printed in Canada.

Fourth printing July 2009.

Paperback ISBN: 978-0-86571-521-9

Inquiries regarding requests to reprint all or part of *Earth-Sheltered House*s should be addressed to New Society Publishers at the address below.

To order directly from the publishers, please call toll-free (North America) 1 (800) 567-6772, or order online at <www.newsociety.com>. Any other inquiries can be directed by mail to:

New Society Publishers
P.O. Box 189, Gabriola Island, BC V0R 1X0, Canada
1 (800) 567-6772

New Society Publishers' mission is to publish books that contribute in fundamental ways to building an ecologically sustainable and just society, and to do so with the least possible impact on the environment, in a manner that models this vision. We are committed to doing this not just through education, but through action. We are acting on our commitment to the world's remaining ancient forests by phasing out our paper supply from ancient forests worldwide. This book is one step toward ending global deforestation and climate change. It is printed on acid-free paper that is **100% old growth forest-free** (100% post-consumer recycled), processed chlorine free, and printed with vegetable-based, low-VOC inks. For further information, or to browse our full list of books and purchase securely, visit our website at: www.newsociety.com.

NEW SOCIETY PUBLISHERS   www.newsociety.com

Lovingly dedicated to Jaki, who has been through it all with me for 33 years: Log End, Earthwood, Mushwood, and, now, Stoneview. Thanks for all that has been and all that is yet to be.

Books for Wiser Living from *Mother Earth News*

Today, more than ever before, our society is seeking ways to live more conscientiously. To help bring you the very best inspiration and information about greener, more sustainable lifestyles, New Society Publishers has joined forces with *Mother Earth News*. For more than 30 years, *Mother Earth News* has been North America's "Original Guide to Living Wisely," creating books and magazines for people with a passion for self-reliance and a desire to live in harmony with nature. Across the countryside and in our cities, New Society Publishers and *Mother Earth News* are leading the way to a wiser, more sustainable world.

Contents

ACKNOWLEDGMENTS

A work such as this is built upon contributions made by others, in this case dating back to the dawn of human history when a perhaps nameless troglodyte began to make improvements to his cave home. Thanks, unknown forebear.

But there are many people I can name who have contributed to my knowledge of earth-sheltered housing or, in some cases, to the actual words of the book. I am honored to list them all amongst my friends: Peter Carpenter, David Woods, and Arthur Quarmby of the British Earth Sheltering Association; underground authors Malcolm Wells and Mike Oehler; owner-builders Geoff Huggins, Richard Guay, Steve Bedard, Jeff Gold, Mark Powers & Mary Hotchkiss, and Chris & Wil Dancey; Ned Doyle for timely help on Appendix A; Dan Rimann, P.E., for reviewing Appendix C; and, of course, my redoubtable editor Richard Freudenberger. It's a great advantage to have an editor knowledgeable in the subject matter itself.

Thanks to Charles D. Nurnberg, president and CEO of Sterling Publishing Inc., (387 Park Avenue South, New York, NY 10016), for his kind permission to use some material which originally appeared in *The Complete Book of Underground Houses* (Sterling, 1994), and *The Complete Book of Cordwood Masonry House-building* (Sterling, 1992).

Thanks to Mari Fox at Colbond Corporation for permission to use illustrations in the colour section and to Marcus Dellinger for supplying Colbond's Enkadrain® drainage composite for field-testing at the Stoneview living roof.

Thanks to New Society publisher – and still friend, I hope – Chris Plant for taking on this project and his patience as I kept missing deadlines, and to Ingrid Witvoet and Greg Green for seeing the project through to completion. Thanks to Diane McIntosh for the cover design.

Thanks to all who contributed pictures to the book. Their names are attached to the captions of the photos themselves.

But primary thanks, as always, must go to my wonderful wife, Jaki, for her patience as the seemingly never-ending work on this book ruffled other parts of our lives, and for her photography, clerical work, and help finding things long lost.

Rob Roy,
Earthwood, West Chazy, New York

INTRODUCTION

Earth-sheltered – or "underground" – housing is one of humankind's earliest shelter options, in the form of natural caves. Sometimes there was competition from other species seeking shelter: mountain lions or bears, for example. As recently as 1987, my friend, earth-sheltered advocate and fellow underground co-conspirator Mike Oehler, had a bit of a territorial skirmish with a large black bear over the occupation of the earth shelter he built and made famous in his classic *The $50 and Up Underground House Book*. Happily, the dispute was amicably resolved.

Mike's ultra low-cost approach to earth-sheltered housing establishes one parameter in the field today. His "PSP" (post, shoring, polythene) method runs the gamut from survival shelter dug out of the ground with "idiot sticks" (pick and shovel) right through some attractive bright comfortable earth shelters making use of inexpensive, often recycled materials. I still sell his book through Earthwood Building School and his building philosophy continues to inspire many people seeking a back-to-nature lifestyle today.

At the opposite parameter of the earth-sheltering literary field is another highly respected friend, the legendary Malcolm (Mac) Wells. Nearly 80 as I write, Mac is no longer active as an architect, but still sells his books out of his Underground Art Gallery in Brewster, Massachusetts. And he remains a wonderful cartoonist. Mac's approach, like Mike's, is highly philosophical, but his construction techniques are from the direction of what I think of as a high-end, powerfully structured and carefully detailed concrete-wrapped approach. Professional earth-sheltering contractors like Marty Davis at Davis Caves in Arlington, Illinois employ structural techniques much more closely related to Mac's methods than Mike's, or my own.

My approach to earth-sheltered construction is at the almost "extreme center" of the field. I guess I've made a niche in the larger field by accenting low to moderate cost methods that can be readily learned and practiced by the owner-builder. I do not intend to depart from that niche in this book, but I will review, revise, and expand the techniques and product information for the present day.

This book came about because of a void left in the field when a previous publisher took two of my earlier books out-of-print in 2003, *The Complete Book of Underground Houses: How to Build a Low-Cost Home* and *Complete Book of Cordwood Masonry Housebuilding: The Earthwood Method*. Although the present work is new and rewritten, a good portion of this book retains the successful techniques described in those earlier works, including the use of surface-bonded block construction, post-and-beam and plank-and-beam roofing, and the use of user-friendly and moderately priced waterproofing and drainage products.

This work will not cover my other favorite building style, cordwood masonry, even though cordwood is one of several good natural building methods that can be employed with above-grade walls of the home. Cordwood masonry is thoroughly covered in *Cordwood Building: The State of the Art*. I will devote a chapter to timber framing (post-and-beam and plank-and-beam) as it pertains to earth sheltered structures, but the subject is more thoroughly covered in my book *Timber Framing for the Rest of Us*. It is my hope that the current work, in combination with my other two New Society titles will provide a complete picture of my building philosophy and methodology, updated for the 21st century.

Our two best-known homes are used as the models for the construction techniques described in this book.

Log End Cave, our first serious foray into "terratecture," could fairly be called an underground home. A couple of years after we sold the Log End homestead, the new owner removed the earth, flattened the ceiling/roof, and built a two-story house on top of the structure. So the underground home, sadly, does not exist in its original form any longer.

Earthwood is a round two-story home with a full earth roof, and with about 40 percent of its cylindrical walls earth-bermed. I call the home an earth shelter, not an underground house. The earth roof and extensive berm, most of it on the northern hemisphere, give the home exceptional energy-efficiency and thermal characteristics. In fact, Earthwood has about 158 percent more usable space than the original Log End Cave (2,400 square feet versus 930 square feet) but uses only 11 percent more fuel (3.33 cords of hardwood versus 3 cords). Why this should be so is discussed in the sidebar. And Log End Cave was by no means inefficient. It, too, stayed comfortable on relatively little fuel.

The building techniques described herein are not limited to the homes used as construction examples. Other shapes and degrees of earth sheltering can be accommodated using these methods. People who want to build their own

Five Reasons for Energy Efficiency at Earthwood

1. *Earth sheltering and earth roof.* The main topics of this book.
2. *Round shape.* Heat loss occurs through a building's exterior fabric. A cylinder encloses 27 percent more space for its skin area than the most efficient rectilinear structure, the square, which hardly anyone builds anymore. All else being equal, energy efficiency is a function of skin area enclosing unit volume.
3. *Cordwood masonry.* Above grade walls are composed of 16-inch-thick cordwood masonry, a building medium that combines excellent insulation and thermal mass in a wonderful way.
4. *Solar orientation.* Earthwood is aligned with the North Star, with most of the double-pane glazing on the southern hemisphere. There is direct solar gain in the winter through at least two of the windows throughout each winter's day (the sunny ones anyway).
5. *The masonry stove.* As at Log End Cave, we heat with wood at Earthwood. However, two-thirds of our fuel goes through the 23-ton masonry stove, which burns wood about 35 percent more efficiently than even the better woodstoves.

While Reason No. 1 is what this book is about, some of the other advantages are discussed in greater detail elsewhere in the book.

house usually want to design it themselves as well, and that is as it should be. My only suggestion to you – for the moment, at any rate – is to keep the structure simple. Simplicity is what will make the house easy to build, economic and, yes, structurally integral. "Keep it Simple" is not necessarily the same as "Keep it Small," although that can be very good advice, too. A small home can be unbearably complicated and a large house can be elegantly simple. What I mean by simplicity of structure will be discussed in the very first chapter, Design.

Chapter 1

EARTH-SHELTERED DESIGN PRINCIPLES

Any discussion of design must begin with an explanation of how earth sheltering and earth roofs can be made to work to our advantage.

There is a popular misconception that earth is a great insulator, and that is why we put houses underground, surround them with earthern berms, or cover them with grass roofs. The reality is that earth is not a very good insulation, with its best insulating characteristics to be found in the first few inches of the soil, where plant roots provide aeration. At depth, where the earth is densely packed, earth is very *poor* as insulation. The misunderstanding of earth as insulation leads to the danger of an equally erroneous view of how earth sheltering really works to provide thermal comfort in a home. If a designer-builder proceeds with an earth-sheltered project from a false understanding of earth's thermal characteristics, the building may not perform well (at best) and could be damp and cold, like so many basements.

EARTH AS THERMAL MASS

Earth's big advantage is as thermal mass. In electronics, a capacitor (or condenser) is a device for storing an electrical charge. I like to think of earth as a capacitor for heat storage. I also like to think of storing "coolth," a word that my word processor tells me does not exist. Coolth is what I call heat at a low temperature. Until you get to absolute zero (defined as the absence of heat) any temperature has a degree of heat to it. To give an example of storing coolth, our 23-ton masonry stove at Earthwood stores heat when it is in use during the winter, but it is also an effective storage medium for storing coolth in the summer. I can still remember back in the 1950s when the iceman delivered blocks of ice around Webster, Massachusetts for use in people's insulated iceboxes. The ice was a capacitor for "storing cold."

So how can we take advantage of this great thermal mass, and not fall subject to potential disadvantage?

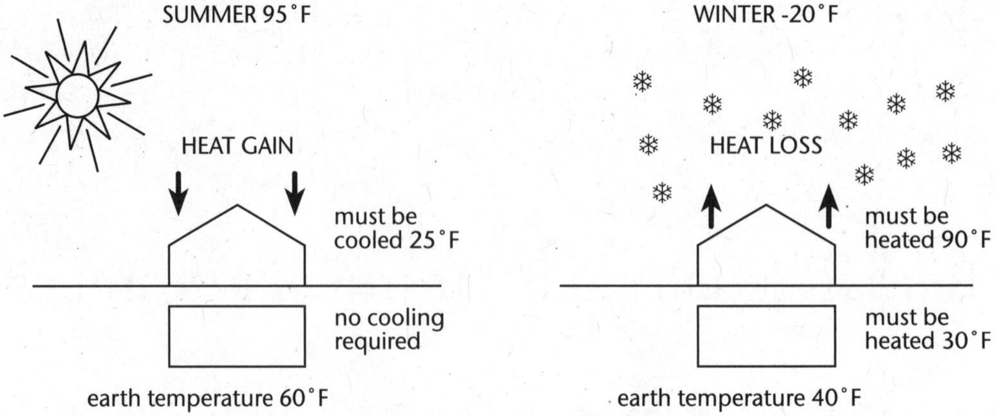

SUMMER 95°F

HEAT GAIN

must be
cooled 25°F

no cooling
required

earth temperature 60°F

WINTER -20°F

HEAT LOSS

must be
heated 90°F

must be
heated 30°F

earth temperature 40°F

Fig. 1.1:
The thermal
advantages of
an earth-
sheltered house,
summer and
winter.

The first ten feet or so below-grade space is a giant thermal mass, which is very slow to change temperature. So there is a real advantage in setting the entire home into this sub-surface climate, which is quite a bit different from the climate above grade. Typically, we are setting a house no more than six to ten feet deep, or "berming" an above grade structure with a similar amount of earth. Our Log End Cave, an "almost underground" home, was set about seven feet below grade, whereas our Earthwood home, built on the surface, is bermed with about 500 tons of earth to a depth of about 13 feet on the northern side, sloping to about 4 feet depth at the southeast and southwest parts of the cylinder.

Figure 1.1 shows summer and winter situations for a house above grade and also for one that is earth-sheltered. The air and earth temperatures given are typical of the range that we would expect to find in the northern United States and southern Canada, from 40 to 50

degrees north latitude. (Pacific coast temperatures might be a little higher, Rocky Mountain temps a little lower.) Our Earthwood home is in northern New York, at 45 degrees north.

The air temperatures range in the example's climate typically varies from about 95 degrees to -20 degrees Fahrenheit, about a 115-degree range. (It can get warmer or colder than these parameters, but rarely.) Note that the temperature range below grade is only about 20 degrees, from about 40 degrees at the beginning of March to about 60 degrees towards the end of August. Temperatures change very slowly below grade, about 1/10 of a degree per day on average. Where we live, a 30- to 40-degree temperature shift from day to night (or day to day) is not uncommon, and we once experienced a change of 70 degrees in a 24-hour period.

Because of its great mass, the earth temperature is slow to respond to climatological changes. This characteristic – sometimes referred to as *thermal lag* – explains why the coldest earth temperature (in the depth range where we typically place earth-sheltered homes) lags about six weeks behind the surface climate, both in winter and in summer. This is why large lakes with huge water masses, like our Lake Champlain, reach their highest water temperatures towards the end of summer, the end of August. It takes all summer to bring the water temperature up to its highest reading. We can take advantage of this thermal lag, both for summertime cooling and wintertime heating.

I think of earth-sheltering a home as the same as building it in a steadier, more favorable climate. Think of how easy it would be to heat and cool a home in a climate that has a range of temperature of 40 to 60 degrees. This is precisely the advantage of earth sheltering. In terms of winter heating, our earth-sheltered home in northern New York performs as if it were built in the coastal plains of the Carolinas. A more favorable ambient temperature in summer yields a similar kind of energy advantage with regard to summertime cooling. In our wintertime situation, the interior of a home on the surface needs to be about 90 degrees warmer than the outside temperature. However, an earth-sheltered home needs to be only 30 degrees warmer than the 40-degree earth temperature on the other side of the wall.

In the summer situation, the surface home in the north needs to be cooled about 25 degrees to achieve comfort level, usually by some energy-expensive air conditioning system. The earth-sheltered home does not require cooling. Nor will it be *too* cool. The earth outside the walls may be at 60 degrees, but residual heat – cooking, sunlight, body temperature, refrigerators, and so forth – will keep the temperature up to comfort level.

Notice that in the commentary above, it is the earth's mass that provides this favorable starting point from which we can begin to heat or cool. We have not begun to bring insulation into the equation. But we must, or we can make a big mistake.

THE IMPORTANCE OF INSULATION

We have spoken of the earth as a thermal mass. It is, in fact, a *huge* thermal mass, and it is not easy to influence its temperature, although it can be done using a rather labor-intensive method called the insulation/watershed umbrella, and described in John Hait's *Passive Annual Heat Storage,* listed in the Bibliography. An insulation "umbrella" extends some distance from the home and encloses the earth near the home. In my view, it is easier and more practical to use the fabric of the building itself as a second and separate thermal mass, one over which we can exercise some control, through the proper placement of insulation. While the home's thermal mass is tiny compared to the earth's, it is still considerable and typically several times greater than the mass of a home built above grade. The Earthwood house, for example, has over 120 tons of thermal mass entirely wrapped inside an insulation barrier. It takes a long time to

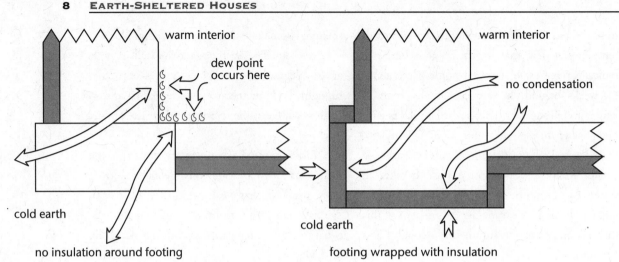

warm interior

dew point
occurs here

cold earth

no insulation around footing

warm interior

no condensation

cold earth

footing wrapped with insulation

change the temperature of 120 tons of something, but it will happen, and, through insulation, we can control the rate of heat transfer, and, thus, the temperature of the mass fabric. Consider the homes in Figure 1.1, wintertime situation. The above-grade home relies on plenty of insulation to protect its inhabitants from the sub-zero outside temperature. The earth-sheltered home also needs insulation, or else the 40-degree earth temperature will actually wick the heat out of the home's fabric through conduction. Without insulation – and *properly placed* insulation at that – the fabric of the building becomes one and the same with the earth's mass. The 120 tons of mass at Earthwood would have the value of a 120-ton slab of stone, a part of the earth itself. Its temperature would be the same as the surrounding earth. Massive walls and floors at 40 degrees would be difficult to heat without insulation.

In order to control the mass fabric of the home itself, we must place the insulation between the home's mass and the earth. In northern climes, we must completely wrap the below-grade portion of the home (concrete, concrete block, stone, etc.) with a layer of substantial insulation. By this method, we use whatever our internal heat source might be (wood, solar, fossil fuels) to "charge up" the mass fabric of the building itself to comfort level. This insulation must be continuous, and without gaps which would create thermal bridges in the mass. I like Mac Wells' term for these: "energy nosebleeds."

I myself made the mistake of not insulating under the footings at Log End Cave, creating a serious energy nosebleed. The left side of Figure 1.2 shows the situation at Log End Cave. Fearing the insulation would be crushed under the great weight of the footings, the 12-inch concrete block walls, and the heavy earth roof they support, I deliberately left out the insulation around the footings. The arrows simply indicate the transfer of heat. Or you can think of it as the transference of coolth; it's all heat at *different* temperatures and, following the law of entropy, doing its best to be the *same* temperature.

Without insulation, conduction through the dense and massive concrete footings causes the inner wall and floor surface near the footings to be about the same temperature as the earth at this depth, say seven or eight feet. Each spring at Log End Cave, particularly in May, June, and early July, when the earth's own mass temperature was still low, any warm moist air created in the home would condense on the cold surface temperature at the base of the external walls, causing condensation, also known as sweating. This is the same effect as you get on the inside of your car windows in the wintertime. Your hot breath condenses on the cold inner surface of the windows. It was late July before the temperature of the footings would get up above dew point, and the sweating would stop. I should have paid better attention to wise Uncle Mac.

At Earthwood, the example on the right, we have not had the problem, because we insulated right around the footings with extruded polystyrene. We have had zero condensation anywhere in the home, and the amount of earth sheltering varies from none on the very south side to four feet at the southeast and southwest parts of the cylinder to 13 feet at the base of the northern part of the earth-sheltered portion of the home. We will discuss the correct insulation to use in Chapter 3.

Any direct conduction, particularly with dense materials such as concrete, metal and stone, can be a serious energy nosebleed. The heat loss isn't even the worst part; the unwanted condensation is the real problem. Always detail some form of thermal break between the house's mass and the earth, or, for that matter, the outside air. We take for granted the importance of continuous insulation above grade. It is at least as important below grade in northern climates.

INSULATION IN THE NORTH

How much insulation should be installed at the various parts of the building's fabric? I think that what we did at Earthwood is a very good pattern for northern climates of 7,000 to 10,000 degree-days. We placed 3 inches of R-5 extruded polystyrene – R-15 total – down to "maximum frost depth" (considered to be about 4 feet in northern New York), and 2 inches (R-10) down to the footings. We went with an inch (R-5) around the footings and under the floor. It should be noted that installers of in-slab radiant heat flooring specify 2 inches of extruded polystyrene – R-10 in all – under the floor.

The situation is a little different in the South, as will be seen below.

This rather lengthy discussion of mass, insulation, and the correct placement of insulation is one of the key concepts of earth sheltering.

Wooly-minded thinking in this matter can cause great problems in the home at the design stage, even before a shovel of earth is moved. So, I'm going to say it one more time, in a slightly different way, with the hope that something twinks with readers who are still unclear on this point: If we do not insulate properly between the home's mass and the earth's mass, we lose our right to easily control the home's temperature. The home would perform thermally like a giant rock set in the earth. Placing the insulation on the interior of the block or concrete wall is a mistake and doesn't help at all with regard to using the home's mass storage potential to our advantage. In fact, an interior insulation would further decrease the temperature of the mass fabric, and if moisture finds its way through this interior insulation, there will be condensation and mold problems which will not be readily accessible, a real catastrophe.

And, too, by insulating on the wrong side of the wall, earth can freeze up against the building and cause structural damage. This occurrence happened to us at the concrete block basement of our first owner-built home, Log End Cottage, when we erroneously insulated on the interior of the block wall. In all of my books, I have made a point of sharing both successes and mistakes. I'm not entirely stupid, so, for the most part, I've learned from my mistakes. But the reader can be more than smart. You can be *wise*, and learn from the mistakes of others, including mine.

INSULATION IN THE SOUTH

As we build further south, insulation strategy should be adjusted, for two reasons. Firstly, at the depths that we set earth-sheltered homes, earth temperatures in the South will be warmer in both summer and winter. With the higher earth temperatures, dew point in the form of sweating is less likely. Second, the further south we go, the more important cooling the home becomes in terms of both comfort and energy cost. By lowering the amount of insulation, the moderating effect of the earth's temperature performs more effectively as a means of natural cooling.

I am from the north and have no personal experience with earth sheltering in the South, but based upon the experiences of people I've spoken with, case studies, and other research, I would suggest an adjustment to insulation placement as per the following two paragraphs. I do not think of myself as the definitive authority on the subject by any means, but am confident that the recommendations below will be an improvement on the insulation recommendations given for northern climes above.

In areas with climates of 4,000 to 7,000 degree days, I would insulate with two inches (R-10) of extruded polystyrene down to the footings, one inch adjacent to and under the footings, and one inch under the floor for a distance of four feet in from the footing. Leave the insulation out from under the rest of the floor to promote natural cooling. If in-floor radiant heat is to be used in a

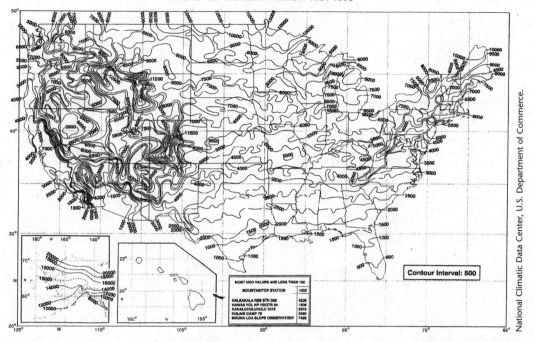

ANNUAL HEATING DEGREE DAYS
BASED ON NORMAL PERIOD 1961-1990

Contour Interval: 500

National Climatic Data Center, U.S. Department of Commerce.

Fig. 1.3: Degree-day map of the United States.

concrete floor, get advice from a local radiant heat installer, who might suggest an inch of insulation under the entire floor, or even two.

In areas of less than 4,000 degree days, insulate with an inch of extruded polystyrene down to the footings, and leave it out around the footings and floor.

Finally, in states bordering the Gulf of Mexico, atmospheric humidity levels can be extremely high, approaching 100 percent a good part of the time. In areas like this, it is almost certain that some form of dehumidification system will be required in the earth-sheltered home. The good news is that earth-sheltering will save a lot of energy on the operating costs of air conditioning.

Figure 1.3 gives a degree-day map of North America (the United States). North of the border area, in Canada, it wouldn't hurt to go with 2 inches (R-10) around the footing and under the floor, if the building budget allows. It'll pay for itself in a few years.

THE EARTH ROOF FACTOR
It is not absolutely necessary to put an earth roof or lightweight living roof on an earth-sheltered house, but not to do so, it seems to me, is a great opportunity lost.

I share seven different advantages to earth roofs with my students, and pull out a timeworn placard to illustrate the points, which are:

1. *Insulation.* I can hear it now: "Hold on there, mate, you just told us earth is poor insulation!" Well, yes, it is. Author and earth-shelter owner-builder Dan Chiras reckons earth is worth about a quarter of an R (.25-R) per inch of thickness, and that rings true with me, particularly for earth a few inches down. But the first three or four inches of earth, where the plant roots aerate the soil, is considerably less dense and, therefore, has some insulation value. The grass or wild-flowers – don't mow 'em – also flop down in the autumn and add more insulation. And, finally, the earth roof holds snow better than any other roof surface, and light fluffy snow is worth a good R-1 per inch of thickness. We notice that our home is even cozier and requires less fuel to heat with a cap of two feet of snow overhead.

2. *Drainage.* With non-earth roof systems, you need some sort of drainage system to remove a lot of water quickly from the roof during a downpour: gutters, downspouts, storm drains, etc. The earth roof drainage – particularly where the roof drains at a single pitch directly onto berms, such as the Log End Cave design – is slow and natural. Even a freestanding earth roof, like the one at Earthwood, must fully saturate before runoff must be attended to.

3. *Aesthetics.* The earth roof is hands-down the most beautiful roof you can put overhead, particularly one of natural wildflowers.

4. *Cooling.* The sun beating down on most roofing causes high surface temperatures. You can literally fry an egg on some of them. The living roof, however, stays nice and cool because of the shading effect of plants, the mass of the earth, and the evaporative cooling effect of stored rainwater. Stick your finger into the living roof and you can feel the coolth.

5. *Longevity.* Built properly, as described in Chapters 7 and 8, the roof will require very little maintenance. We don't even mow ours anymore. All other roofs are

Fig. 1.4: Author Rob Roy and his timeworn placard.

Dan Strickland

subject to deterioration from the ultraviolet (UV) rays of the sun, from wind and water erosion, and from something called freeze-thaw cycling. In our climate near Montreal, most roofs are subjected to between 30 and 35 freeze-thaw cycles each winter, and each occurrence breaks the roofing down on the molecular level. Sun, wind, and frost never get to the roofing surface, so, protected by the earth from these adverse conditions, the waterproofing membrane is virtually non-biodegrad-able. It should last 100 years or forever, whichever comes first.

6. *Ecology*. While not the right place to grow shrubs, trees or root vegetables, the earth roof can support all sorts of plants and microbial life. Instead of killing off – say – 1,500 square feet of the planet's surface to yet more hot, lifeless black tarscape, we can return the home's footprint to cool green oxygenating living production. We'll discuss vegetation options in detail in Chapter 8.

7. *Protection*. Just a few inches of earth afford all sorts of protections not found with other roofing surfaces: fire, radiation, and sound, just to name three. In combination with a Log End Cave-type berm, the earth roof can also contribute to tornado, hurricane, and earthquake protection, as well.

The seven advantages to an earth roof all occur with just a few inches of earth on the roof. Doubling the thickness, from – say – 6 inches to 12 inches does not double the value of the advantage. With fire and sound protection, for example, extra earth beyond six inches adds little advantage; you've still got windows, doors and some portion of above grade walls influencing these considerations. But doubling the earth does double the potential saturated load of the earth component of the roof. And this extra load greatly increases the structural cost of the home. I accent timber framing (also called "plank and beam" roofing) as the most suitable roof structural system for the inexperienced owner-builder. Other options, such as pre-stressed concrete planks (which are very expensive and must be installed with a crane), or poured-in-place reinforced concrete roofing (which should be professionally installed), not only add greatly to the structural cost, but also don't look nearly so nice overhead as a ceiling.

Fig. 1.5: Wildflowers in bloom atop the Earthwood office.

Roy's General Theory on Earth or Living Roofs

This is as good a time as any to tell you about the realities of the earth roof with regard to its weight, or load. Calculating the desired load is the first step towards designing.

But, first, a clarification of terms might be in order. I use the term "earth roof" to describe a roof system that relies primarily on a certain thickness of earth or topsoil to nurture the desired vegetation or ground cover. A "living roof" might have earth on it, or it might have some other growing medium for the plants, such as straw. Or it might combine earth with another medium, as we have done on the roof of our straw bale guesthouse. Lots of work has been done in the past ten years in both Europe and North America on these alternatives that eliminate or considerably diminish the need for placing heavy earth overhead. And the reason is usually an effort to keep the structural cost down. I will share some of the methodologies in use in Chapters 7 and 8.

Because of the heavy weight of saturated earth, my theory for 25 years has always been – and still remains – that we want to use enough earth to maintain the green cover, and not a whole lot more. The reason, as we will soon see, is that wet earth is very heavy, and a great depth of it – while technically possible – adds unacceptably to the structural cost for owner-builders who want to own the home themselves, and not have a bank own it for them.

At Log End Cave and at Earthwood, we had good success with maintaining a green cover (mowed grass at the Cave, wild at Earthwood) with an earth roof with a final compacted depth of about six inches of soil. A couple of times at Earthwood, during its so-far 23-year life, the roof almost died off during drought. We never watered it. But always, after some compensating rain, the roof would come back and flourish once again.

Back in the 1970s, several builders were placing from 18 inches to 3 feet of earth on the roof, and, yes, they engineered the structure properly to support that kind of load. But these homes were very expensive, with a good part of the expense caught up in the roof support system. Why they did it remains a little unclear to me. After you've got an honest 6 inches of earth on the roof, the seven advantages listed above are present. If additional insulation is desired, earth is a poor choice. An extra inch of extruded polystyrene, which weighs practically nothing, will yield as much additional R-value as an extra foot of earth. And neither do the other advantages listed increase proportionally to the use of greater amounts of earth.

How Much Does it Weigh?

Now, let's "do the numbers" as they say on a popular National Public Radio program. Essentially, there are two different loads that need to be added together to arrive at the grand humungous total for which the building must be engineered: the dead load and the live load. The dead load is sometimes called the structural load, and refers to the fabric of the building itself: the rafters, the planking, the waterproofing membrane, insulation, the drainage layer, and the like. Firstly, any building we design has to be able to support itself. Over the next few paragraphs, I'll be referring to Table 1 (p. 17), which lists the weight or load of various common materials.

The Earthwood dead load consists of 5-inch-by-10-inch red pine rafters, 2-inch-by-6-inch tongue-in-groove planking, the Bituthene® 3000 waterproofing membrane, and four inches of Dow Blueboard™ insulation (an extruded polystyrene). Everything above the Styrofoam® we'll put under the classification of live load, as its weight is subject to change. The dead load calculations are on page 16.

I've put everything else, even the crushed stone and earth, under the classification of live load (also on p. 16), because all of these components vary with conditions.

This is kind of a "worse case" scenario, and that is what we have to engineer for. Note that most of the load is live. If you have difficulty picturing 150 psf, think of 1,400 people packed in on a 1,400 square foot roof (Earthwood with overhang). Each person averages 150 pounds in weight and each occupies a square foot. People are live loads, too, but it is unlikely that they will be up on the roof at the same time as the maximum snow load.

Now, a couple of points must be made here. The first is that in life's reality, just when you think that things can't get any worse, they do. I have been told by a reliable source that the ski resort of Whistler, British Columbia, must engineer roofs for a snow and ice load of 350 pounds per square foot. They get several feet of snow, then rain and a freeze, then more snow, then more rain and freeze, and several feet of very dense snow and ice can accumulate on the roof. And these are not even earth roofs. So Plattsburgh's 70 pound snow load, while rare, could, under extreme conditions, be even worse. This is why people physically (and wisely) shovel excess snow off their roofs during extreme snowstorms, especially anyone living in a mobile home. Note also that with the 6-inch-thick earth roof at Earthwood, almost half of the total maximum structural load is the snow load, so find out what this load is for your area, and keep in mind the earth roof holds snow better than any other kind. This is a positive characteristic in terms of insulation, but a possible negative with regard to load.

The second point is that when engineers do their stress load calculations carefully and accurately, using the right formulas and the right unit stress numbers for the grade and species of wooden rafters and girders (or metal beams or pre-stressed concrete planks or whatever), they will have built in a safety factor of approximately six. This is not overbuilt. This is the kind of

Dead Loads

Rafters: A 5-inch-by-10-inch red pine rafter weighs about 8.4 pounds per linear foot. Although it is a radial rafter system, it is fair, for purposes of load calculation, to figure an average spacing of 24 inches center to center, designated as 24 inches o.c. If rafters were 12 inches o.c., a linear foot of rafter would support a square foot of roof, and the rafters themselves would add 8.4 pounds per square foot (8.4 psf). However, at 24 inches o.c, each rafter supports 2 square feet, so the load per square foot would be exactly half, or 4.2 psf.

4.2 psf

Planking: We used 2-inch-by-6-inch tongue-in-groove planking, which weighs six pounds per square foot. To get this figure, I actually weighed some planks, and divided by the area of the planks weighed.

3.0 psf

W.R. Grace Bituthene® 3000 waterproofing membrane: (I weighed a square foot of it on a postal scale.)

0.4 psf

4 inches of Dow Styrofoam® Blueboard™: at 2 pounds per cubic foot density.

0.7 psf

Dead Load Total   **8.3 psf**

Live Loads

2 inches crushed stone: at 100 pounds per cubic foot.

16.7 psf

6 inches of fully saturated earth: at 108 pounds per cubic foot (from engineering charts and actual measurement by the author).

54.0 psf

3 inches loose hay or straw: filtration mat, matted down and wet. Estimated.

1.0 psf

Snow load: for our area, as per Plattsburgh, NY building regulations.

70.0 psf

Live Load Total   **141.7 psf**
Combined Dead and Live Load   **150.0 psf**

Table 1
Weights/Loads of Some Common Materials

WEIGHTS IN POUNDS PER CUBIC FOOT (pcf)		WEIGHTS IN POUNDS PER SQUARE FOOT (psf)	
Clay, wet	114.0	Bituthene®	0.38
Concrete	150.0	Enkadrain®	0.24
Crushed stone (#2)	111.0	Plywood, (¾")	2.40
Earth, saturated	108.0	Rafters (4 × 8 pine, 12" o.c.)	4.60
Granite	165.0	Rafters (5 × 10 pine, 12" o.c.)	8.40
Pine, yellow (dry)	45.9	Rafters (5 × 10 pine, 24" o.c.)	4.20
Oak, red (dry)	44.0	Spruce planking (2 × 6, T&G)	3.00
Oak, live (dry)	59.0	Styrofoam® (2")	0.33
Sand, wet	120.0		
Sand, wet and packed	130.0		
Sawdust	17.0		
Water	62.4		

Left column is in pounds per cubic foot (pcf). Right column is in pounds per square foot (psf). To obtain pounds per inch of thickness per square foot of area, divide numbers in the left column by 12. Example: An inch of concrete weighs 12.5 pounds per square foot because 150 ÷ 12 = 12.5. And four inches of concrete would weigh 50 pounds per square foot because 12.5 × 4 = 50.

"redundancy" that is considered to be good engineering.

If an engineer "works to the numbers," the building should be safe. But nature can be cruel. The power pylons in Montreal were designed to support 1.75 inches of ice. But the ice storm of 1998 put over 4 inches of ice on the steel pylons. The pylons folded up and most of the power grid supply into the city failed. Another famous example of nature defeating engineering is the collapse of the Hartford Civic Center roof under a severe ice and snow load in 1978. Fortunately, 5,000 hockey fans had vacated the building just a few hours before the collapse. Not so fortunately, and germane to the topic of this book, author and owner-builder advocate Ken Kern died when an earth roof he built collapsed upon him. From anecdotal evidence from intermediary contacts, it is my impression that his building was not properly engineered; it failed under heavy mud conditions.

I mention these things not to scare the reader – Kern was, perhaps, pushing the envelope in his design and it is a singular case, to my knowledge – but rather to impress upon you the importance of working to the numbers and having the work checked by a qualified structural engineer. Many jurisdictions may require an engineer's stamp on the plans. If these steps are taken, the earth roof is no more dangerous than any other kind. In fact, with regard to tornadoes, forest fires, and – yes – radiation, they are, in my view, safer.

Calculating load is the easy part, using Table 1, but a necessary first step.

I will be discuss the design considerations of timber framing in Chapter 6, but the subject is more completely covered in my earlier work, *Timber Framing for the Rest of Us*, particularly Chapter 2, "Basic Timber Frame Structure," and Appendix B, "Stress Load Calculations for Beams." This appendix, repeated in this book as Appendix C, shows how to "work to the numbers" for the 40-by-40-foot Log End Cave plan. However, the formulas and methods are useful for any timber frame roofing system having parallel rafters.

SIMPLICITY OF DESIGN

The reality of most home designs today, whether they are architect-designed and contractor-built, owner-built (and designed), earth-sheltered or plunked on the surface, is that they are difficult to envision, engineer, draw and build. All of which leads to complications, more expense than necessary, and more time to build the thing. With the architect-designed home, which is usually professionally built, this is not a huge problem, particularly for the architect and the builder, both of whom benefit financially from a more complicated (read *expensive*) home.

But I am going to assume that the reader has bought this book with a view to saving money, perhaps even avoiding mortgage altogether. So we need to talk about simplicity of design, as a great deal of cost savings can occur at the design stage.

Maintain uniformity of structure, style and line. Keep the structural design itself as simple as possible. Let's consider a home with two intersecting rooflines, as, for example, a regular gable-ended two-pitched roof intersecting with a similar roof at right angles to the first. The intersection creates alternating ridges and valleys, a tricky detail. I hear the refrain: "But, lots of houses are built this way." True, but almost always by professional contractors, used to doing this sort of thing day in and day out. Fortunately, earth-roofed homes should have a shallow-pitched roof anyway, which begs a simple roof line for valuable structural reasons, a single-pitched shed roof, for example, or two simple shallow pitches meeting at a ridge, such as the basic Log End Cave design featured as a design model in succeeding chapters.

Round houses fall under the classification of simple structures. After all, the other building species on this planet almost always employ a round design, as do the so-called "primitive" human builders in Africa and Asia. The Earthwood round house, for example, is simple because of the radial rafter system. There are sixteen identical roof facets repeating one after the other until you go around the building, back to the point of beginning. All of this simplicity comes from the initial installation of – say – 16 rafters, or other simply divisible number.

However, as simple as the round shape is to design and build, there is plenty of room to complicate the design abysmally by the addition of rectilinear rooms. Trying to marry the radial rafter system of the round part with parallel rafters in the rectilinear addition is a nightmare of design and of building. Round houses are not easy to add onto for this reason, although we did so successfully at Earthwood, by extending the rafters of two of the sixteen roof facets outward. If such an addition is anticipated (by rafter extension), no matter what the house shape, it should be accommodated for at the design stage. We were fortunate at Earthwood that the overhanging rafters were high enough that I could continue outward from the original round building the extra eight feet for our new sun room and still maintain sufficient ceiling height in the new roof. The detailed construction of this addition, from the design stage through to the earth roof, makes up the entire fifth and final

chapter of my companion work, *Timber Framing for the Rest of Us.*

Different room sizes, particularly when heavy earth roofs are part of the equation, will normally require different structural designs. However, engineering and construction can be simplified by keeping the rafter (ceiling) spans the same as much as possible. The length of the roof is not so critical, as rafter spans in earth-roofed building will normally span the room's width. So a longer room simply requires more rafters. Frequency (spacing) and rafter dimensions can stay the same, as long as span stays the same. The room could be a hundred feet long.

Usually, people arrive at complicated structural designs because they start out with a floor plan, not a structural plan. They know what they want in terms of layout, but haven't got the foggiest idea of how to build it. They take a piece of paper – some have the presence of mind to make it a piece of graph paper with quarter-inch light blue lines intersected in a quarter-inch grid – and then proceed to draw rooms where they would like them. If a simple roof framing plan is not considered at this very early stage, the result is often a complicated structural plan. *This is particularly important given the structural demands of an earth roof.* Therefore …

Integrate the floor plan with the structural plan. In fact, I start most building projects with a simple structural plan and try to make the design elements of my idealized floor plan integrate with that structural plan. I'm going to show how this works a little later on with the two distinct models used in this book, Log End Cave and Earthwood.

Does this integration necessitate compromises in the floor plan? Yes, it does. Is it worth it? Yes, it is, particularly if (1) you want to get the house built in a reasonable amount of time, and (2) you want to own it yourself instead of the bank owning it for you.

Choose an efficient shape. I have said that round is a simple shape to build, and I'll back up this statement later on by showing how to build it in a simple way. But round is also an efficient shape in terms of enclosing the maximum amount of space with the least amount of perimeter wall material. An added advantage of this favorable relationship (most area per unit perimeter) is that the round house has the least skin area to translate into heat loss. Earth sheltering reduces – but does not eliminate – heat loss, so minimizing skin area is still a good thing, as it is with any home.

Other shapes can be quite efficient, too, particularly with a passive solar earth-sheltered home. The square is actually the most efficient of the rectilinear designs with regard to enclosing the most space with the least materials. Richard Guay (and others) have built the 40-by-40-foot Log End Cave plan given on pages 27 and 28 for this reason. However, it is not bad to go longer in the east-west dimension than the north-south dimension, particularly with an earth-sheltered passive solar design. This elongated rectilinear design was quite popular in the heyday of earth-sheltered housing (back in the 1980s) and with good reason: it facilitated the use of more south-facing windows for passive solar gain and light. Decreasing the depth of the house into the earth meant that natural sunlight – particularly the low-lying winter sunbeams – would penetrate deeper into the home. With earth-sheltered or underground housing, we take every opportunity to maximize natural light. So, a small house of – say – 20 feet deep by 40 or 50 feet long might make sense, or a medium-sized house of 30 feet deep by 40 to 60 feet long might work for a larger family. (The word "deep" in this paragraph refers to the depth of the building back into the hillside or berm, not the depth below grade.) Our original Log End Cave was roughly 30 feet deep by 35 feet long and the winter sun penetrated right to the back of the house.

Efficient shapes can be combined with a simple modular design for ease and economy of construction. I like using 10-foot rafter spans as the standard in rectilinear construction. The rooms don't have to be 10 feet square, although they can be. Any length is possible while still maintaining the simple structural plan based upon 10-foot spans. We'll be talking a lot about span in this book, but a 10-foot span is one that enables the owner-builder to use reasonably sized timbers that are moderate in cost and fairly easy to handle. Greater spans can be used, but, beyond 12 feet, the cost and weight of the timbers become a very serious consideration.

Keep it small. In and of itself, building a smaller house does not necessarily equate with building a simple house. I have seen students bring small but impossibly complicated designs to consultation sessions at Earthwood Building School. An example was a nautilus-shaped home with no two rafters the same size or length and tremendous waste of linear or sheet materials, and very difficult and time-consuming details. The house would have been quite nice if completed, but it was going to be a long, drawn-out, and probably expensive proposition for an inexperienced owner-builder. "Small is beautiful," says E.F. Schumacher. Well, the design was small and beautiful, but its chances of completion by any but a very crafty and skilled builder were next to nil.

Types of Earth-Sheltered Design

1. True Underground

I have only visited one truly underground home in the United States. In fact, it is the only one I've even heard of. It was John Barnard's Ecology House in Marstons Mills, Massachusetts. It used to be open to the public, but Mac Wells told me recently that an above-grade home has built on top of it, relegating it to the same fate as our original Log End Cave. Residents accessed the home by descending a stairway into an open below-grade courtyard. The various rooms of the home opened onto this courtyard and were also interconnected with each other under the earth roof that covered the entire interior space.

Fig. 1.6: *Heavy timbers support a 15" to 18" thick earth roof at Baldtop Dugout in New Hampshire, designed by Don Metz.*

In the excellent but out-of-print *Earth Sheltered Housing Design*, we learn that in the arid regions of eastern central China, a deep loess soil provides "ideal conditions for self-supporting excavations" below grade. Loess is a wind-driven sandstone, soft enough that it can be carved with a hand-adze, but strong enough to maintain structural integrity. The book says that "millions of these cave dwellings" have been constructed through to the present day. "In many locations, farming continues on the surface above the below-grade houses. Approximately 20 million people live in cave dwellings in China today."

In the late Ken Kern's book, *The Owner-Built Earth Shelter*, we learn of the wonderful underground home hand-carved over a 40-year period into Fresno, California's hardpan strata by one Baldasare Forestiere (1879–1946), a Sicilian

Fig. 1.7: Here's an old view of John Barnard's Ecology House on Cape Cod, a rare American example of a truly underground home. The space now serves as a basement beneath a new surface home.

immigrant: "Forestiere chiseled from unpromising land a subterranean environment for himself, fruit trees, and vines. Using only his hands and a pickaxe, shovel, and wheelbarrow, he invested time, perseverance, and hard work to realize his dream. In the process, he created an intriguing architectural paradigm for modern times: a strong, easy to build, low-cost underground habitat suitable for plants and people. To this day, his structure remains entirely functional, environmentally harmonious, and incomparably beautiful."

All told, over a ten-acre site, Forestiere created rooms, tunnels, gardens and underground patios as a home and an orchard. On one citrus tree, he grafted seven different fruits. The site, operated by the Forestiere family, is open to the public. Go to <www.undergroundgardens.com> for more information and pictures.

2. Bermed

Most homes with their external walls earth-sheltered are not really "underground." Most have an earthen berm placed against the structure. Or, the building might be set partially below grade, with the excavated material used to berm the above-grade portion. In the 1970s, a study made by the University of Minnesota's Underground Space Center found that berming a home can have 90 to 95 percent of the energy advantage as placing it totally below grade. The berming option, in my view, makes for an easier home to build, and one which is almost certainly lower in cost than the home build fully below grade.

Also, meeting egress code (see page 24) is easier with the bermed home, by the use of what are called penetrational entrances. Figure 1.8 is a good example of an entranceway penetrating a berm. Windows can also be penetrational, as per Figure 1.9 below.

With the bermed style, there is very often a fair amount of exterior wall above grade, as was true at both Log End Cave and Earthwood, where we used cordwood masonry walls above grade. In northern climates, these above-grade exterior walls will generally fall on the south or southwest part of the house to maximize solar gain and minimize heat loss.

3. Atrium

Quite a few homes built in the halcyon days of earth-sheltered housing – roughly 1970 to 1985 –

made use of atriums or courtyards, often towards the side of the house opposite to an exposed south elevation. These atriums could be covered with glass to provide a pleasant indoor/outdoor space, or they could be left as a Spanish style open courtyard, as seen in Figure 1.7. Whether they were glass-covered or open, they let light into rooms that might otherwise be short of light, and they could provide a route to a second means of escape to meet code.

4. Two Stories?

My guess would be that a majority of earth-sheltered houses are single story. In fact, in the Underground Space Center books *Earth Sheltered Housing Design* and *Earth-Sheltered Homes* (both, sadly, now long out of print) a quick tally of 44 homes reveals that only about a quarter of earth-shelters are two stories. Log End Cave was one story and Earthwood has two.

A flat or gently sloped site suggests one story, where a more steeply sloped site might work well with two stories, particularly if the slope is facing the right direction for your energy concerns. Earthwood was built on a flat excavated gravel pit set about five feet below original grade. We tried to tuck the house up close to the edge of the gravel pit to take advantage of that five feet of elevation, but we still did a lot of berming, probably close to 500 tons of material, to build the berm up to its present 13 feet of height on the north.

A word of caution here: A steep slope should not be dug into with impunity. It may be that ground cover is what is keeping that slope stable. The type of subsoil is critical, too. Is it sandy, gravelly, loamy, or clay-like? The "angle of repose" – the slope at which a pile of material can support itself – is important. Before excavation – in fact, before *design* – you'd be well-advised to have the

Fig. 1.8 (left): Earthtech 5, a Don Metz design, features a good example of an entrance-way penetrating a berm.

Fig. 1.9 (right): Arthur Quarmby's home in Yorkshire, England, has the master bedroom window penetrating the earth berm.

site examined by a qualified soils engineer. He or she can also help you with important soils and percolation information with regard to designing a septic system on the site.

Finally, after you have determined that you can safely build into a particular hillside, it is still prudent to design a "curtain drain" or surface drain to carry runoff away from the home. I'll say it a lot in this book: Drainage is always the better part of waterproofing.

EGRESS AND FENESTRATION CODE CONSIDERATIONS

With any and all of the various styles of earth-sheltered housing described above, it is important to meet code with regard to egress in case of fire or other emergency, such as the code enforcement officer showing up at the front door. In short, every public room of any dwelling must have two separate means of escape, including bedrooms.

At first, earth-sheltering might seem to be an impediment against providing the required egress, but, with a little thought – and sometimes a little inconvenience – it is not difficult to meet code. Penetrational doorways through berms can allow just as many exits – and as sensibly-placed – as any above-ground home, even though the retaining walls required might add somewhat to the time and cost.

Another provision of the code is worth knowing about, as it can be used advantageously. Opening windows can satisfy egress code requirements under the International Residential Code (and others) providing that they are positioned no more than 42 inches (sometimes 44 inches) off the floor and have a clear "escape and rescue opening" with a "minimum net clearance opening" of not less than 4.0 square feet (International Building Code), although the details of this provision vary from code to code and year to year. A penetrational window in a bedroom, for example, might be an alternative to the more costly and troublesome penetrational door. The codebooks are very difficult to interpret, so run your plans by the very code enforcement officer who is responsible for issuing your building permit. There are at least four different codes used in the various states, and Canadian provinces have different requirements as well.

Finally, clever floorplan design can be employed to allow two separate escape paths from the various "public rooms." Say you've got a master bedroom with an attached bath. There could be a second door out of the bath leading to a different exterior door than the main bedroom entrance would normally use.

Fenestration codes (minimal amount of glazing required) may or may not be as strictly enforced as egress requirements, but you should know about them. Codes still vary from state to state, and they are not the easiest reading in the world, which is why it is good to have your plans checked by a professional. Remember that windows can take a variety of forms: penetrational through the berm, elevational on unbermed sides (such as sliding glass doors), skylights, and windows or glass doors into atriums.

A little thought is all that is necessary on egress and fenestration issues, but the important thing is to be aware at the design stage, because you know that your local code official is watching for it at the approval stage. The code manual being used in your area will be necessary to supplement my generalized overview here.

The Original Log End Cave Plan

I show this plan (Figures 1.10, 1.11 and 1.12) for historical interest (about 1984, Log End Cave became a basement under a new two-story home), and to show how a floor plan can integrate with a structural plan. The space described by the structure can be divided in different ways, according to individual needs. The "sauna" shown on the plan was used exactly once as such, and became a storage room. (We built a free-standing earth-

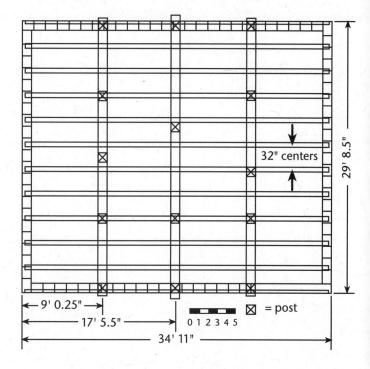

roofed sauna midway between the Cottage and the Cave.)

We liked the open-plan living/kitchen/dining "great room" with its two centrally located

Fig. 1.10: Block, rafter and timber frame plan for the original Log End Cave.

Fig. 1.11: South wall elevation at the original Log End Cave.

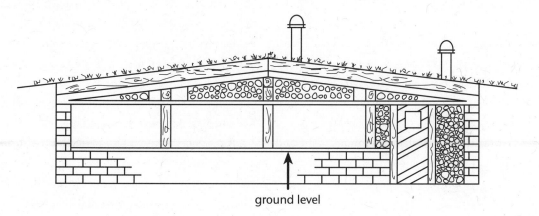

ground level

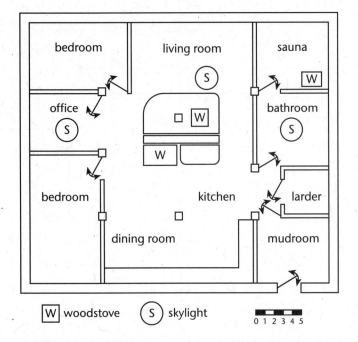

bedroom

living room

Ⓢ

sauna

W

office

Ⓢ

□ W

bathroom

Ⓢ

W

bedroom

kitchen

larder

□

dining room

mudroom

W woodstove   Ⓢ skylight   0 1 2 3 4 5

Fig. 1.12:
Log End Cave
floor plan.

THE 40-BY-40' LOG END CAVE PLAN

Many people also wanted greater spans in the rooms, so I designed the 40-by-40-foot Log End Cave plan shown in Figures 1.13, 1.14 and 1.15. This basic plan, with minor space use changes, has been built several times around the country, including Richard and Lisa Guay's home (see Figures 2.2 and 2.3).

Room width spans are almost 10 feet, and the home has three separate means of escape to facilitate compliance with egress code. Before building, always have your floor plans checked for code-worthiness by your local building inspector.

THE EARTHWOOD PLAN

Birds, bees, beavers and Bantu tribesmen in Africa all know instinctively, and without benefit of a course in geometry, that the most efficient use of materials and labor to enclose a given space is the circular house. But if you are curious about the geometry, I can tell you that the walls of a round house enclose about 27 percent more space than the most efficient of the rectilinear shapes, which is the square (a shape hardly anyone builds anymore, although it used to be popular precisely because it was efficient), and 42 percent more efficient than the more commonly selected shape, a rectangle about twice as long as it is wide.

A two-story house is also more compact than a single story. Compact, in this case, means cheaper to build and more energy-efficient. There's only one foundation and one roof (the

woodstoves. We could regulate the temperature in the peripheral rooms by leaving doors open or closed.

The 30-by-35-foot shape was an efficient use of materials to enclose space, with the longer dimension facing south for solar gain.

The obvious flaw with the original Log End Cave plans is that it does not meet code with regard to ingress and egress. But my friend and cordwood masonry compatriot, Jack Henstridge, reassures me: "None of us need feel totally useless, Rob. We can always serve as horrible examples."

At least it was structurally sound, light and bright, and comfortable to live in.

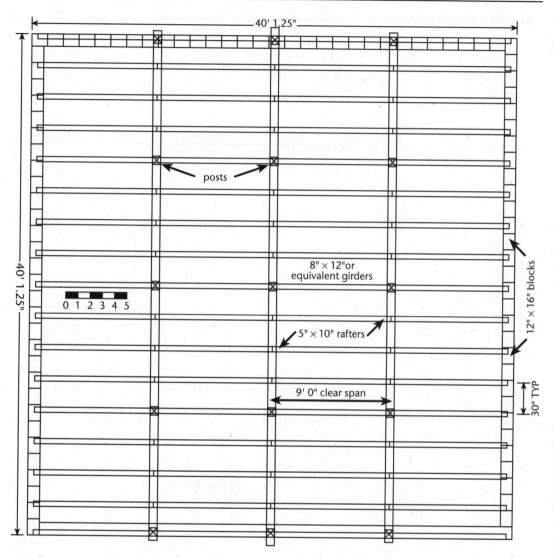

posts

8" × 12"or
equivalent girders

12" × 16" blocks

5" × 10" rafters

9' 0" clear span

30" TYP

40' 1.25"

40' 1.25"

0 1 2 3 4 5

Fig. 1.13:
Block, rafter and
timber frame
plan for the
40 × 40' Log
End Cave.

expensive surfaces, compared with the intermediate floor), but the space of the home is doubled.

Our floor plan is given in Figure 1.16, and the structural plan in Fig. 1.17. I place them together here to illustrate how the structural and floor plans integrate with each other, just as they did with the Cave plans. How an individual actually uses space is up to them, but these plans might plant some seeds of thought.

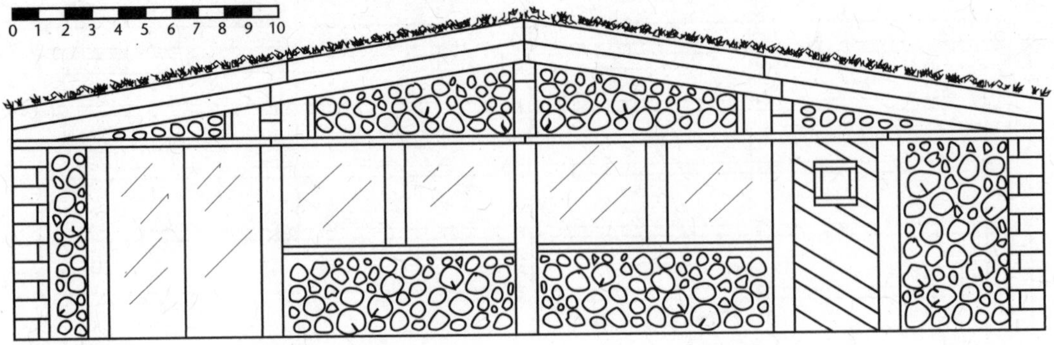

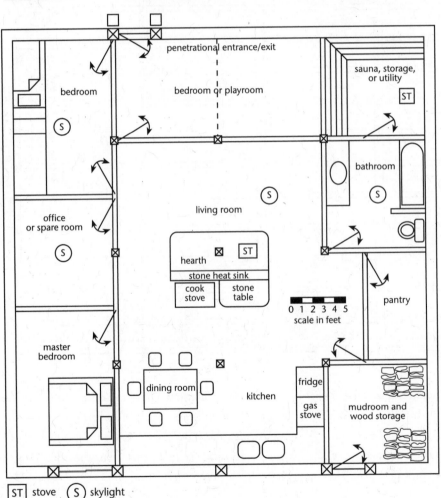

Fig. 1.14:
South wall
elevation for the
40 × 40' Log
End Cave.

Fig. 1.15:
The 40 × 40'
floor plan can
be adjusted to
meet individual
space use
requirements,
but keep in
mind the
10 × 10' square
modules upon
which the
engineering is
based.

penetrational entrance/exit

sauna, storage,
or utility

bedroom

bedroom or playroom

bathroom

office
or spare room

living room

hearth

stone heat sink

cook
stove

stone
table

0 1 2 3 4 5
scale in feet

pantry

master
bedroom

dining room

kitchen

fridge

gas
stove

mudroom and
wood storage

ST stove   S skylight

The Evolution of Design at Earthwood

I wrote these words in 1983 for the long out-of-print *Earthwood: Building Low-Cost Alternative Houses*. I still stand by them:

A house design is best allowed to evolve at its own rate. I do not intend to be mystical when I say that much of this evolutionary process takes place in the subconscious, in dreams, perhaps, or during those few lucid moments between wakefulness and sleep. And many problems of design seem to solve themselves, although I expect that what really happens is that passing time enables the designer to get a new perspective. Flash-in-the-pan special features which occupied the imagination so intently at their inception are seen in a more practical light three months down the road. Changes in thinking occur almost imperceptibly. I cannot even recall when Jaki and I first made the decision that Earthwood would be our home. Perhaps there was no decision, just a realization. But I do know this: when we left for Scotland just before Christmas, 1980, for a five-week visit, we knew not only that the house was for us, but that it would be a two-story round house with two other nearby round outbuildings.

The Earthwood structural plan calls for a post-and-beam frame midway between the central mass and the exterior walls, in order to cut the rafter and floor joist spans in half. An octagon would work nicely here, but I suffer from a lifelong fixation with poking colored plastic balls around a large green table, and a billiard table will not fit in the Earthwood plan if I chose an octagonal post-and-beam frame. But it would fit if I could eliminate just one of the posts, effectively truncating the octagon, something I was able to accomplish with a 14-foot-4-inch-long full-sized 9-by-12-inch oak girder spanning from, say, Post 6 to Post 8, thus eliminating Post 7 of the octagon. Pillar footings for these seven posts are discussed in Chapter 3: Footings.

My father used to say you had to build two houses to get one right, one to make all your mistakes in. Well, my father was cleverer than me. It has taken us three houses to get one just the way we like it, and Earthwood is that house.

That doesn't mean we won't build another, though.

USING "MARGINAL" LAND

As owning my own home is preferable to the bank owning it for me, I'm constantly looking for ways to save major chunks of change. Mac Wells

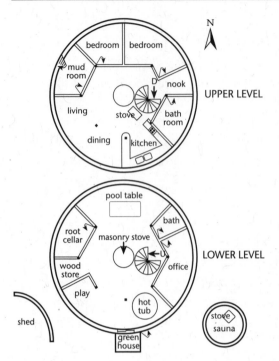

UPPER LEVEL

LOWER LEVEL

Fig. 1.16: The original floor plan at Earthwood can be adjusted to meet individual space use requirements, but it is advantageous – structurally and economically – to integrate the floor plan with structural plan.

woodland had been taken down five feet for the removal of gravel back in the 1950s and 60s.

When we bought the property, not a thing was growing on this gravel pit, not even weeds. There were four surrounding acres of good woods around the pit, however. Still, this great gaping crater in the middle of the woods made the property "dirt cheap." Now, 23 years after building Earthwood, visitors are surprised to hear that the house site was once a moonscape. We have reclaimed nearly two acres of the planet's dead surface and returned it to living oxygenating food-producing greenscape. It's a good feeling to do some small part in reversing the sorry trend of development in this country.

Sometimes loggers "clear cut" a woodlot. The more considerate ones leave a little for regeneration. The logging greatly reduces the value of the property, and it is amazing how fast the trees come back. With a little care, you can create a wonderful forest. Usually, seedlings and healthy small trees are already there to give the process a jumpstart.

Mac built his first underground office building in a dump in Cherry Hill, New Jersey, the ultimate land reclamation project. You get the idea.

DESIGN AROUND BARGAINS

The previous section describes great savings that can come from the use of reclaimed land. A similar kind of economic advantage can come from designing around recycled or other bargain materials. These materials might be literally at

inspired me to look for "marginal land," which might be clear-cut woods, an old rubbish tip, a fire-killed forest, a gravel pit or simply land which is low in cost because of poor access or distance from commercial electric.

Any house that we can design and build is, to some degree, an imposition on nature. The earth-sheltered house has the potential for the least negative impact, because we can return the house site's footprint on the planet back to its original earth-covered state. But why chew up nature and try to restore it, when we can use land, which has already been chewed up, and then restore it to a better condition than whence we found it? This is what we did at Earthwood. Almost two acres of

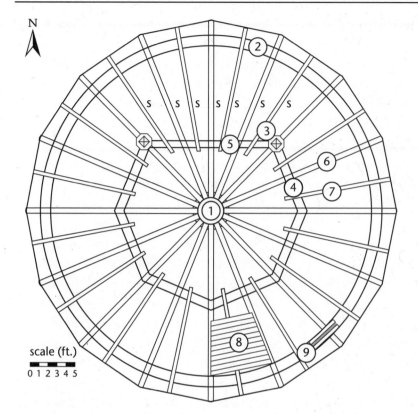

N

scale (ft.)

0 1 2 3 4 5

Fig. 1.17:

The Earthwood structural plan.

1. *Stone heat sink or masonry stove, 4' diameter upstairs, 5' diameter downstairs*
2. *16" cordwood masonry wall above grade, 16" block wall below grade*
3. *Post locations, seven in number*
4. *Girders, eight-by-eight best Douglas fir or equal*
5. *Ten-by-twelve clear oak girder or equal*
6. *Primary rafters (five-by-ten best white pine or equal) or floor joists (four-by-eight)*
7. *Secondary rafters, as per 6 above.*
8. *Two-by-six tongue-in-groove planking*
9. *Two-by-six plates, typical*
S = *Special larger rafters for greater spans, six-by-ten best white pine or equal. Floor joists at these locations can be made of four-by-eight material.*

your feet. Our "worthless" gravel pit supplied us with all the sand we needed for Earthwood, enough good building stone to build our 23-ton masonry stove, and plenty of material to berm up the northern hemisphere of the home. My son, Rohan, recently bought a couple of acres in southwestern Colorado. The crushed shale soil, it turns out, makes an excellent cob material for building external walls. Trees, obviously, can yield posts and beams, even lumber. And trees unsuitable for making into

lumber might be perfectly good for cordwood masonry construction.

And that's just indigenous materials.

Many cities have Habitat for Humanity stores, which sell a tremendous variety of perfectly good recycled building materials. Jaki and I looked at one in Durango with Rohan and were amazed at the bargains on excellent windows, doors, cabinets, even plumbing and electric fixtures.

Don't keep it a secret that you are need of building materials and they will almost miraculously come your way. The author James Redfield (*The Celestine Prophecy*) speaks in terms of "cultivating coincidences." It works. Listen:

Just a couple of months ago, I was trying to explain the science of coincidence creation to a young couple who were thinking of building a home. We were walking along the dirt road where we live with another young couple, mutual friends, who had just built a home. I said, "Look, I'll demonstrate how this works. Jaki and I need a shower cabinet for the guesthouse we're working on at Earthwood. A decent shower cabinet can cost two or three hundred bucks. Last month, I mentioned this to friends in Quebec and it turned out that they have an extra one they're not using, and I can have it just for the hauling. Trouble is, it's up in Quebec. But I am mentioning my need to you folks, as well, because you might know of one closer by."

I was trying to make a point, but it got better, real fast. It turned out that both couples that I was walking with knew of shower cabinets that I could get nearby! "There," I said, matter-of-factly, "now I've got a choice of three shower cabinets." Everyone laughed and shook their heads like I had just done something magical and mystical. But I hadn't. I'd simply set the laws of probability into motion, in my favor.

I'll trouble the reader with just one more small example, as it may save you 100 times the cost of this book. Every decent-sized population center has a manufacturer of thermalpane glass windows. Some cities have several such firms.

Invariably, these companies have a "back room," sometimes known as the "pig pen," where they store perfectly good thermalpane units which, for one reason or another, were never picked up by the customer. There may have been a bankruptcy; maybe the units were made the wrong size, whatever. The nature of glass is that it is cheaper for the manufacturer to make a new unit from large sheets of virgin glass than it is to take these miscut or uncollected units apart, clean the sealant off the edges, and make a smaller unit from them. It is also very troublesome. They're simply not set up to do that. You come along and buy the units – *perfectly good, no flaws* – for ten to twenty cents on the dollar. Before we designed Earthwood, we bought twenty clear one-inch thick fixed thermalpane units, a quarter-inch plate glass each side, with a half-inch space of sealed inert gas between. They were all 15½ inches by 42½ inches in size, each a half-inch too big to fit in the rough openings of a new bank in town. We bought the lot for $75. Well, they were perfect for the round Earthwood house.

You design around bargains like that.

PLANS AND MODELS

You've got your land. You know where to get good bargains on building materials. You've figured out a structural plan that is simple to build and accommodates your desired use of space, and some of the key materials you've scored. Now you can draw a detailed plan, at quarter-inch or half-inch to the foot. Large sheets of graph paper can be useful for maintaining scale. Use a pencil that has, on its other end, the most valuable tool you own: an eraser. I like to figure out the structure first (with the rough floor plan always in the back of my mind) and then tweak it to make the floor plan fit in a tidy manner. You will want a plan showing the location and thickness of the external walls (block, poured, stone, cordwood, whatever), the post and girt locations, floor joists, and roof rafters. You can make copies of this and work in the floor plan, adjusting both the structural and floor plans as necessary to make them mesh together. You will also want elevations for the various sides of the house, and perhaps a typical section view through the middle of the structure. These drawings help you to visualize the actual construction.

Jack Henstridge, ever a font of homespun wisdom, says: "Build a model. You'll learn a lot about the actual construction that way. If you can't build the model, for Pete's sake, don't try to build the house." It's true. Sometimes, a difficult jointing detail is hard to visualize on flat paper, but you can work it out in the scale model. You can have a lot of fun – and learn a lot about building – by making a model. They make doll furniture it a scale of an inch to the foot, a nice scale to work with. A 30-by-40-foot design will be fairly large, though, on the kitchen table: 2-foot 6-inches by 3-foot 4-inches.

Sidewall elevations are almost meaningless on a round house, and difficult to draw, because the curvature recedes away from the vantage point.

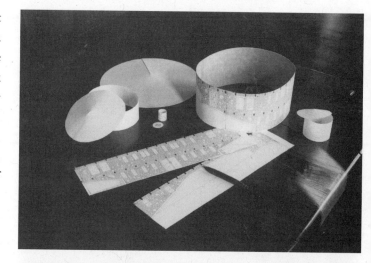

Therefore, I made a three- dimensional plan at a quarter-inch to the foot, so that I could work out all of my window and door placement with respect to the rafters, floor joists and other structural components. This can be seen in Figures 1.18 and 1.19. A roof with the right pitch was easy to make, which gave a really good idea of what the house would finally look like.

Figs. 1.18 & 1.19: It is useful – and quite easy – to make a paper model of a round house.

Cost is very important to me. But so is aesthetics.

If this turns out to be the longest chapter in this book, then so be it. That's appropriate. The design stage is where you – *and that eraser* – can save a lot of physical work and money later on. I spent about five weeks working out the plans for Earthwood. It took us seven months to close the building in to the point where we could heat it. It's a big house and I like to think that careful planning allowed us to beat the winter during our all-too-short North Country building season.

So, with plans in hand, let's move on to construction.

Chapter 2

SITING & EXCAVATION

Because an earth-sheltered house really needs to be site specific, siting considerations need to be done at the design stage. But a discussion of siting also ties in well with excavation, so we'll do it here. Besides, Chapter 1 was long enough already.

Earthwood siting was dead simple. The land was already as level as a parking lot and we wanted to place the home on the north edge of this gravel pit, to take advantage of full solar exposure to the south. Our job was simply to build the house, nice and strong, and then berm it with 500 tons of earth. To lessen earth-moving somewhat, we did take advantage of the 5-foot-higher grade on the undisturbed north edge of the pit.

But the siting at Log End Cave was probably a more typical example, certainly more interesting and more instructive. We wanted to set the Cave into a gentle south-facing slope at Log End Homestead, the ideal site, really, and one that underground enthusiasts dream about.

First, let's work with earth volumes on a simple flat site excavation, because the math is a lot easier. It always surprises (and sometimes delights) one or two of my earth-sheltered housing students to learn that they can do an earth-sheltered house on a perfectly flat site, provided that they are not digging into the water table. With low-lying flat land, the following example should be avoided. If the site suffers from lack of drainage, you can't be sure of keeping the water out of your earth-sheltered house even with the best of membranes. In short, don't build down into the water table. How do you know? Do a 6-foot "deep-hole test" with a backhoe on the site in question. A qualified soils engineer will be able to tell you where high water table is (along with other useful information), even though you excavate in a dry season. Chances are you will need a deep-hole test anyway, to meet health department requirements for septic system design approval. This is a New York State requirement now, although it is not always enforced in some of the outlying areas.

Wastewater Systems

At the earliest stages of design, careful consideration must be given to the disposal of wastewater. Composting toilets and other alternative systems which make use of "humanure" are admirable, and we have used some of these ideas successfully at Earthwood. An excellent book on the subject, *Humanure* by Joe Jenkins is listed in the Bibliography. However, in New York State, and many others, even though composting systems are allowed, they must be accompanied by a state-approved conventional septic system. Silly? Probably.

Conventional systems based on pumps are expensive beyond the economic viewpoint of this book. And pumping systems are subject to continued maintenance, constant power consumption, reliability problems, and failure if and when the power goes out. Gravity, on the other hand, works for us day in and day out.

So integrate the siting of your septic tank and drainfield very carefully with the site of the earth-sheltered house itself. On a gentle slope, waste drainage by gravity won't be a problem. On a flat site, you might need to dig deeper tracks for the drainfield than normal, or keep the elevation of the house up higher (which will mean more berming), or even raise the level of the bathroom by a few inches to establish the correct gradient for a gravity system.

In other words, you can build below grade providing you can drain a frog-strangling downpour away from the site. This will require either excellent percolation of soils down to a deep water table, or enough slope to take a French drain out above grade, even if it means a hundred-foot track. Don't try to live inside an inside-out swimming pool. The water will beat you eventually.

A flat or nearly flat site can make good use of a semi-bermed design. As we will see, a relatively shallow excavation of just 30 inches or so can yield enough material to berm the three non-elevational sides. Of prime importance in keeping costs down is to minimize bringing material in ... or hauling it away. Let's do the numbers, based on the Log End Cave type of design shown in Figure 2.1.

FLAT SITE EXCAVATION CALCULATIONS

The topsoil is going to go back on the roof or the berm, and needn't really affect the calculations very much. Typically, there might be six to eight inches of organic material – grass and topsoil – that should be scraped to the edge of the site and mounded into piles for use on top of the home at its completion. If your topsoil layer is very thin, or of very poor quality, you might want to consider bringing in an extra truckload of good stuff.

For our example, we'll assume that we scrape six inches of soil to the edge of the site, handy for later use, but out of the way. A bulldozer is the best piece of equipment for the job, but a good backhoe operator can do it, too, although it might take a little longer. While a backhoe is the more common machine used for excavation, the bulldozer can actually do a good job in ground conditions free of large rocks. Topsoil should be removed not only from the house footprint, but also from the bermed area, as well, typically 15 feet (5 yards) all around the building.

With the topsoil out of the way, we're down to what architects and engineers call "undisturbed earth," or, sometimes, "the clay layer," even though it may not be clay at all.

The house plan used as an example is 30 feet square, but we'll convert everything to yards (10 yards square) because cubic yards is the usual unit used for excavation. The diagram is in yards, too. You can work in feet if you like; just divide cubic feet by 27 to get cubic yards. We'll do this later for concrete calculations, too.

East and west sidewalls are, say, six feet high, with rafters above. It's a small building. As the ceiling pitches up to a ridgepole in the center, the building has plenty of headroom. The north side

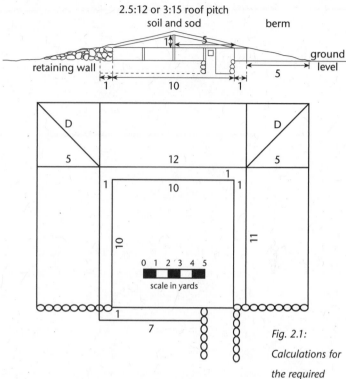

Fig. 2.1:

Calculations for the required excavation depth for this gabled berm-style 30 × 30' house should take into account whether or not the excavated material will be suitable for backfilling against the sidewalls.

of the home will be bermed to the seven foot level, but the south elevation will not be bermed at all, in favor of plenty of thermalpane glass windows for solar gain.

Just for fun, we'll assume that the subsoil is of poor percolating characteristics (often the case), necessitating the hauling in of good percolating backfill, such as coarse sand, for placing against the house walls. (An alternative strategy, using manufactured drainage material, will be discussed in Chapter 7.)

I suppose calculus could be used to advantage, but not by me. I like "calculated guessing" myself. (Isn't that what "calculus" means, anyway?) Let's

Energy Saving Tip

Every square foot of south facing double-pane insulated glass, in combination with thermal curtains or shutters closed at night, will be a net energy gain into the home. Triple pane insulated glass, because of reflection, is not as good. We say it has a lower rate of insolation. Single pane glass would actually let the most solar heat in, but the heavy heat loss at night would more than offset the gain. So, on the south side, in northern climes, double pane insulated glass, at about R-2, is best. Particularly if you install, and use, insulated curtains, as we do at Earthwood.

say that we excavate one yard (36 inches) deep over all the area within a yard of the house wall. The resulting square to be excavated will be 12 yards on a side, or 144 square yards. As it is a yard deep, the excavation will yield 144 *cubic* yards. Because our hypothetical site has poor percolating soils, the yard of space won't be backfilled with the same material that comes out of the hole. We'll bring in sand for that purpose. How much earth will the rest of the berm require?

The three sides of the berm directly next to that necessary yard of sand backfill has the cross-section of a right triangle: one yard high (h) and 5 yards wide (b); that is, a cross-sectional area of 2.5 square yards. (A = ½ bh = ½ × 5 × 1 = 2.5). The total length of the berm is 34 yards (11 + 12 + 11) where it is directly adjacent to the sand backfill, so the volume is 2½ times 34, or 85 cubic yards. To this, we must add the volume of the two delta-wing shapes where the berms meet at the NE and NW corners, marked "D" in Figure 2.1. The

volume formula for these corners is ¼ hb², so, by substitution, ¼ × 1 × 5 × 5 = 6.25 cubic yards. In all, it will require 97.5 cubic yards of earth to build the berm (85 + 6.25 + 6.25 = 97.5). But we've taken 144 cubic yards out of the hole. The 46.5 cubic yard difference is quite a bit to haul away or spread around.

Before we make a second "calculated guess," let's pretend we're building in nice percolating sandy soil, instead of poor percolating material. The yard of backfill up against the east, west, and north walls can be thought of as a rectangular volume measuring a yard wide, two yards high, and 32 yards long (10 + 1 + 10 + 1 + 10 = 32). The volume for a rectilinear solid (V = lwh) results in 64 cubic yards (1 × 2 × 32 = 64). Now, the total volume of the berm right up to the house walls is 161.5 cubic yards (97.5, from the previous paragraph, + 64, the backfill within a yard of the house, = 161.5), a bit more than the 144 cubic yards that came out of the hole. But this is a ballpark calculation. The berms could be made a little steeper, or the excavation made a couple of inches deeper, to make things come out very nicely. A skilled operator will grade the area wisely with the available material.

But let's go back to the poor soil example for a second guess. This time, let's just excavate 30 inches (2.5 feet, .833 yard) instead of 3 feet. Now, the volume of the excavation will be 0.833 of what it had been before (because 2.5 feet divided by 3 feet = 0.833). The excavation yields 120 cubic yards of earth (0.833 × 144 = 120). The sand used for good drainage near the building's walls will remain the same at 64 cubic yards. The berm is six inches higher now. Also, the 2.5-in-12 roof pitch (expressed 2.5:12) established by the roof adds 30 inches to the width of the berm at original ground level.

Note: Roof pitch is conventionally measured as units of rise per twelve of run, inches, feet, whatever. In Figure 2.1 (above), I've drawn the peak one yard (3 feet) higher than the east and west sidewalls, which are 15 feet away. This establishes the 2.5:12 roof pitch. An earth roof should have a pitch of at least 1:12 to promote drainage, but not more than about 3:12 to prevent the earth from slumping towards the edge of the building. The original Log End Cave pitch was 1.75:12, while Earthwood has done well with a pitch of just over 1:12. My friend Mike Oehler will go as high as a 5:12 pitch, but Mike and I agree to disagree on this.

Expressed in yards, then, the berm is now 1.17 yards high and 5.83 yards wide. The volume of the

berm is ½ bhl (where l = length taken along the inside of the berm) plus 2 × (¼ hb²) for the two delta wing shapes marked "D." By substitution: (.5)(5.83)(1.17)(34) + (2)(.25)(1.17)(5.83)(5.83) = 115.72 + 19.85 = 135.57 cubic yards. Thank goodness for calculators. We dug 120 cubic yards out of the hole. Not bad. A bit deeper than 2½ feet (30 inches) should be just about right.

The examples might seem a bit tedious to the computationally disinclined, but they are realistic.

Fig. 2.2: Richard Guay's 40 × 40' Log End Cave, during construction.

Fig. 2.3: Richard and Lisa Guay's completed earth-sheltered home in Champlain, New York. The solar room addition was paid for with some of Lisa's winnings as a five-time Jeopardy champion.

(The obvious design fault of the single entrance, a mistake we actually made on the original Log End Cave in 1977, can be remedied by some sort of penetrational entryway on the north side (see Figure 1.8) or even a second door on the south side, as former student Richard Guay did at his 40-by-40-foot Log End Cave built in 1980. (See Figures 2.2 and 2.3.)

Such a shallow excavation of – say – 30 inches might not suggest "underground housing" to the reader, but it performs almost as well as a home totally below original grade. From a distance, the house would look like a knoll on the landscape.

Author's Note: From this point on in the book, I will be describing the various building techniques as they have applied to the two major earth shelters we have built. Some headings cite "Log End Cave," and the information in those sections is appropriate for rectilinear structures, or excavated sites. Other section headings have "Earthwood" in their titles, with the text more directed to the special considerations of a round house and/or the floating slab. Where neither home is cited in a heading, the text can be assumed to be useful with any earth shelter.

LOG END CAVE: "A GENTLE SOUTH-FACING SLOPE"

The very words are music to the ears of an earth-sheltered advocate – one living in the North, at least. But, with regard to earth-moving, a slope complicates the math somewhat. At Log End Cave, Jaki and I found it much easier to visualize

the whole project once we created an accurate contour map of the immediate area. It was easy to do and very worthwhile. Here's how we did it:

At the top of the knoll at Log End Homestead, we set up a transit (loaned to us by a surveyor friend). A contractor's level, or laser level, available quite inexpensively at equipment rent-all stores, will accomplish the same thing. We plumbed and leveled the transit to a point on the ground within the legs of the tripod, and marked the spot with a half-brick sitting on a large red-tagged nail stuck in the ground.

We drove a wooden stake along the true north-south alignment (the meridian), maybe forty feet south of our benchmark nail. To find south, get a morning paper or watch the evening weather report, each of which usually tell the time of sunrise and sunset for that day. Halfway between those two times, the sun will be at true south. Or use a compass, but be sure to add or subtract the magnetic declination for your area, available from the US Geological Survey data or from a local airport.

We set the zero-degree mark of the transit's compass rose to a fixed point (the corner of the existing Log End Cottage) in case we needed to reset. I stayed with the transit and Jaki would walk away from me with one end of a 50-foot tape in one hand and a calibrated grade stick in the other. You can hire such a stick at the rent-all store or make one yourself by marking up a two-by-four.

Our procedure was to establish a slope for each ray of the circle divisible by 15 degrees: 0°,

15°, 30°, 45°, and so on. I'd set the transit at 0 degrees, for example, and Jaki would move away from me with the grade stick until we were able to discern a one-inch drop in the land. We then measured from the benchmark to Jaki's grade stick, and recorded the distance on a clipboard. Then Jaki would move further away along the same ray until we read a 6-inch drop. Again, we measured and recorded the distance. We recorded every 6-inch drop of elevation along the ray until we were outside the immediate vicinity of the house site. We repeated this procedure along the 15-degree ray, and so on, right to the 180-degree ray. We now had a statistical abstract of the building site, seen in Table 2. The project took us most of an afternoon.

I put the transit away and transcribed the abstract onto a large piece of graph paper having a grid of five squares to the inch. I let each square equal a foot and drew a light pencil line at each 15 degrees of arc for the half-circle in which the house would fall. Then, with a ruler, I measured the scale distances along each arc and placed a pencil dot at the appropriate distance for each recorded drop in elevation. To avoid confusion, I labeled each dot lightly: -1", -6", -12" and so on.

We considered the minus one-inch contour to be level for our purposes. When all the figures from the statistical abstract were transcribed to the graph paper, I connected dots of like elevation with gently curving lines. A very accurate contour map emerged as the dots were connected, something like the center drawing of Figure 2.4, below.

The time spent on the contour map was more than made up by eliminating guesswork later on. Using a piece of the same graph paper used for the contour map, I made a scale model of Log End Cave's foundation plan. Then, siting was a simple matter of sliding the little square of paper around the contour map until the most sensible location emerged.

Our plan called for the tops of the east, west and north block walls to be 78 inches above the concrete floor slab. The door on the south side would enter straight to the exterior, which would slope slightly away from the home. And we decided to berm up to the underside of the three large windows on the south side, something I wouldn't do again, as snow begins to accumulate in front of the windows. The top of the 36-inch-high block wall under the windows would be 42 inches below the top of the other three walls. You can see at the top of Figure 2.4 that the north and south walls are about even with original grade. In fact, the south wall did require about an extra 6 inches of excavation.

The major amount of the excavated earth is used to build up the east and west berms. This way, the new ground level melds nicely into the shallow-pitched earth-covered roof.

We also had to factor in the delivery of over 20 loads of sand to backfill the home, since the soils at Log End had very poor percolation, as we found out to our cost at the Log End Cottage basement. The site's contour map and the little foundation model made this all possible.

Table 2
Statistical Abstract of Land Contours (Original Log End Cave Site)

*CONTOUR	-90°	-75°	-60°	-45°	-30°	-15°	0°
1"	23'7"	16'5"	12'2"	9'4"	7'7"	6'8"	6'1"
6"	33'4"	29'4"	24'9"	21'6"	18'6"	15'11"	15'
12"	36'	34'2"	32'8"	27'11"	22'7"	21"	19'4"
18"			36'	31'10"	27'11"	26'	24'2"
24"				36'1"	33'	30'2"	27'10"
30"				41'4"	37'	32'4"	30'4"
36"				46'4"	40'10"	38'2"	35'2"
42"					47'	42'5"	40'6"
48"					54'	49'2"	46'7"

*CONTOUR	+15°	+30°	+45°	+60°	+75°	+90°
1"	6'	6'3"	6'6"	7'	7'8"	8'
6"	15'7"	16'2"	17'2"	16'7"	17'	21'6"
12"	22'6"	21'7"	20'11"	21'10"	27'2"	29'
18"	24'4"	23'	22'5"	26'3"	31'4"	35'8"
24"	26'6"	25'10"	25'11"	30'1"	36'	40'8"
30"	19'9"	29'1"	29'4"	34'	38'10"	45'3"
36"	34'8"	32'8"	33'7"	37'5"	42'4"	49'7"
42"	38'2"	36'8"	36'3"	39'5"	46'4"	
48"	43'1"	40'10"	40'1"	41'9"		

*Contours in inches below reference point.

(Measurements indicate the distance from the reference point to certain contours along primary rays.)

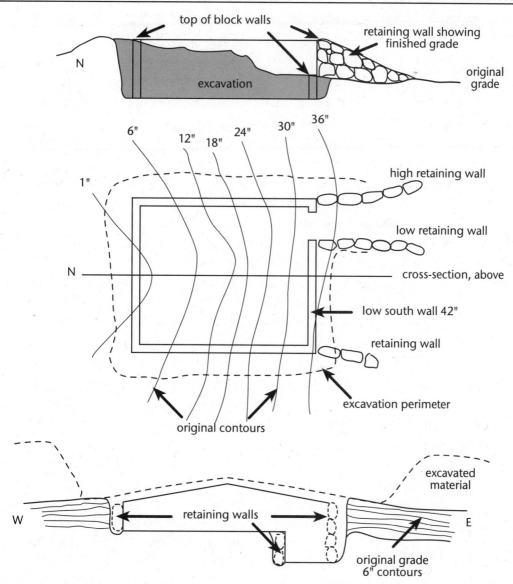

top of block walls

retaining wall showing
finished grade

N

excavation

original
grade

6"

1"   12"   18"   24"   30"   36"

high retaining wall

low retaining wall

N   cross-section, above

low south wall 42"

retaining wall

excavation perimeter

original contours

excavated
material

W   retaining walls   E

original grade
6" contours

Fig. 2.4:
The north and
south walls of
Log End Cave
were almost
even with the
original ground
level. The east
and west wall
berms were built
up using
excavated
material.

Landscaping Log End Cave at the completion of construction went very well indeed.

Incidentally, I'd never build a basement again. If you are tempted to go with a "basement," I implore you to think beyond basement to proper light, bright, dry, warm earth-sheltered space, which is what this book is all about. Go the extra mile with waterproofing, drainage, ventilation,

Mike Oehler on Basements

Underground co-conspirator Mike Oehler says, in Chapter 1 of his *$50 and Up Underground House Book*: "An underground house has no more in common with a basement than a penthouse apartment has in common with a hot, dark, dusty attic."

Further, on page 9, he goes on to say: "A basement is not designed for human habitat. It is a place to put the furnace and store junk. It is constructed to reach below the frost line so that the frost heaves don't crumple the fragile conventional structure above. It is a place where workmen can walk around checking for termites under the flooring, where they may work on pipes and wiring. ... A basement is usually a dark, damp, dirty place and even when it is not, even when it is a recreation room, say, it is usually an airless place with few windows, artificially lighted and having an artificial feel. An underground house is not this at all."

and layout strategies that are not commonly employed with basements. Yes, this all adds to the cost, but you'll have a livable space instead of a dark, damp, dismal … basement. See Mike Oehler's sidebar.

The site's contour map and the little foundation model made this all possible. Landscaping Log End Cave at the completion of construction went very well indeed.

LOG END CAVE: LAYING OUT THE EXCAVATION

With my foundation plan sensibly placed on the contour map, it was a simple matter to set the transit up again and place the four corners of the foundation on the ground for the edification of the backhoe operator. It's a simple angle-and-distance plotting. For example, with a small plastic protractor and a ruler, it was easy to learn from the map that the northeast corner of the foundation should be 18-feet-3-inches from our benchmark, at an angle of 29 degrees. We transposed the angle and measurement at full scale on the site, marked the spot with a wooden stake, and did the same for the other three corners. The foundation corners would soon be lost during excavation, however, so we used a simple method of setting four additional stakes four feet out from the foundation stakes, illustrated in Figure 2.5. I had a four-foot-square half sheet of plywood lying around and used it as shown to establish four more stakes, which would indicate the corners of the excavation itself. Now the front-end loader operator could visualize both the foundation and the excavation parameters.

As the poor percolating soils had a fairly steep angle of repose, he was able to create a flat base at the proper depth, with a good 30 inches of working room all around the foundation, just right. Our contour map could be used to explain how deep to go at each of the four corners.

Figures 2.6 and 2.7 tell a little about angle of repose and creating sufficient working space around the building. Eventually, because of poor percolating soils, we would fill this space from the gravel pit at Earthwood, which we owned at the time, just a half-mile away. With sandy soils and a shallow angle of repose, as in Figure 2.7, it will be necessary to create a larger excavation, with the outer stakes six feet or even eight feet out from the foundation stakes. But, on the positive side, with good percolating sandy soils you will be able to backfill with the same material that comes out of the hole.

Again, a deep-hole test, useful for septic system design in the area, can teach you a lot about excavation considerations: water table, angle of repose, and more. Unexpected ledge or hardpan might change the whole project strategy. Maybe the good sandy soil is only three feet deep, with clay below. It is nice to know these things.

"Can the removed earth be used to backfill the home?"

This is an important question. If percolation throughout the excavation is good, as with sandy or gravelly soils, then the answer is affirmative. But if the earth doesn't let water through, as with

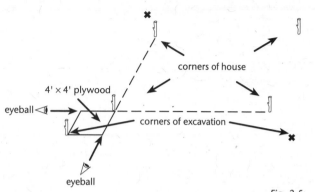

Fig. 2.5: *Establishing the corners of the excavation at Log End Cave.*

clay soils, then the answer is no. A good indication would be puddles that take forever to percolate away, or only disappear through evaporation. With poor percolating soils, you will need to bring in good backfill, such as sand, or use one of the drainage products described in Chapter 7. Your choice of strategies in this matter will probably be economically driven, weighing the cost of hauling in sand versus the cost of the drainage matting.

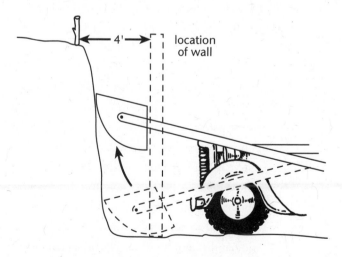

Fig. 2.6: A four-foot work space all around the footing is plenty of space for building the walls.

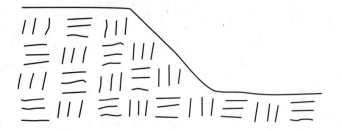

If you've got poor drainage, it's probably a good idea to read Chapter 7 now. If you've got really horrendous drainage, such as soil designated as "expansive clay" or if the deep hole test reveals water table three feet down, you'd better rethink your plans altogether. Maybe you can berm a house built much nearer the surface or switch to an earth-roofed above-grade home. Underground housing does not suit every site.

Radon ...

... is a clear odorless gas, which, in sufficient quantity, can cause lung cancer and other health problems. It can enter underground homes through cracks in the foundation or other openings. As an underground home is largely below grade, the risks can be quite a bit higher than with a house on the surface. If your area is renowned for high incidence of radon, or if you are building in gravel over shale, granite, or phosphate deposits, it would be a good idea to test your soils for radon before proceeding. This can be done in the deep hole test, for example, but there is inexpensive equipment available to enable you to conduct a radon test for your site without the deep-hole test. The presence of radon does not absolutely disqualify the site for earth-sheltered housing, but the techniques which make it safe to build there are expensive, and require the use of special venting materials around the foundation to exhaust radon gas out to the atmosphere. Personally, I'd rethink the project if heavy radon is present. See Appendix A for more information, including what concentrations are considered safe.

Excavation Costs

Combined with the ensuing backfilling and landscaping, excavation is one of the biggest and costliest jobs associated with earth-sheltered housing.

Unless you are experienced in operating heavy equipment yourself or have some masochistic desire to employ "idiot sticks" (picks and shovels), you will need a heavy equipment contractor. Get at least three estimates. You can get estimates for the whole excavation, or you can pay the contractor by the hour for the equipment. If the contractor is experienced, the job estimate should be pretty close to what it would cost on a per-hour basis. However – and I have asked contractor friends about this – they may build in a "contingency percentage" as part of their "complete job" estimate, to take into account all the various imponderables that can happen on site: large boulders, etc. Contractors have told me that the customer, therefore, is usually better hiring by the hour, and this is the way I do it. With a new contractor or one you're not experienced with, it's a good idea to be on site and check on the hours. Logging mistakes can be made, and, at $50 an hour (or more), they can amount to quite a bit. Normal practice is to charge one-way haulage of the equipment to your site. The next customer pays the next delivery charge. With remote sites, you may have to swallow a haulage charge in both directions, but find this out ahead of time to avoid nasty surprises.

Compare apples to apples. Let's say Contractor A gets $50 an hour for his backhoe, while Contractor B charges $60. "A" uses an 18-inch bucket, while B's bucket is 30 inches. For excavation, B might be less expensive, other things being equal. For a drainfield, where you don't necessarily want 30-inch tracks (which need to be filled with expensive crushed stone), Contractor A might be the better choice.

Remember that you'll need the equipment back later on for backfilling, landscaping, maybe the septic system. I have found that contractors are inclined to treat your back pocket better if you give them a volume of work. I stayed with the same contractor, Ed Garrow, throughout the Log End Cave project. I liked his work so much that I retained him at Earthwood, too. Ed even built our megalithic stone circle for us a few years later, and enjoyed doing it. I get excellent service as a regular customer.

Price is important, but so is ability and reputation. If two contractors give you similar estimates, but one has a better reputation for quality, go with the good rep, even if it costs a few dollars more. You want to be happy with the way they leave the site, for example, which can save you hours of handwork later on.

Another reason that I like paying by the hour – instead of the job – is that there are so many imponderables in earth-sheltered housing. You might change your plans a little, such as deciding where to put a gray water soakaway (dry well), for example. If you are paying by the job, the contractor might penalize you for changes, and rightfully so.

Which Kind of Equipment is Best?

A client of mine in Georgia is building an earth-sheltered house as I write this book, early 2005. He had never operated mechanized equipment before, but decided to hire a "Bobcat," which is a small, stable, wheeled loader. In an hour or so, he learned to operate the machine to a pretty fair degree of competence, and proceeded to excavate the entire site for his small 24-by-36-foot earth shelter in a few days. He saved a lot of money, got the excavation exactly the way he wanted it, and had fun in the bargain.

At Log End Cottage, we had our cellar hole dug with a backhoe, as it was all below relatively level terrain. This worked well, so I assumed that I would need a backhoe for the Cave, as well. But Ed Garrow suggested a front-end loader. The loader was $22 an hour (these are 1977 dollars; you can triple the numbers today!) and the backhoe was $16, but the loader did the job in about half the time as the backhoe, so it was the better deal.

Why was the loader faster? Unlike the Cottage basement, the Cave excavation was cut into a hillside. It was an easy matter for the loader, with its six-foot bucket, to maneuver. The home's south elevational entrance was the natural way for the loader to come in and out of the excavation easily. True, the loader has to back away with each bucketful and dump it, while a backhoe can stay in one spot for a while and, with its long articulated bucket, place the earth outside of the excavation. But the loader moves about a cubic yard – or more - with each scoop.

In point of fact, when the loader was beginning to travel rather far with each load, we did finish some of the corners with a backhoe, which we needed anyway to dig the footing tracks (Chapter 3). It's an advantage to hire a contractor with a variety of equipment. Make sure you are only paying for one at a time, though.

The front-end loader would also do well for the excavation of the bermed house used as an example earlier in this chapter. Remember, too, that in sandy soils, a good operator on a bulldozer would do a good job quickly. In the first scrape, the topsoil could be pushed into a pile. With the blade set deeper for each pass, the dozer could strategically pile the excavated earth for ease of backfilling the bermed home later on.

Whatever equipment you use, prepare a flat site which extends at least three feet beyond the outside edge of your planned foundation.

A backhoe is best for digging septic lines, drain fields and any other needed ditches, but normally you'd wait until the house construction is finished before doing these jobs. You want to be sure that grades are right, and, too, you don't want a lot of dangerous holes and ditches around the work site.

EARTHWOOD: SITE WORK

The typical approach to American housing is to tear up the land to moonscape, build the house, and then attempt to reclaim the earth through ubiquitous "landscaping." At Earthwood's site, in a gravel pit, the first step had already been taken for us. We could proceed with a minimum of site preparation. There were no trees or any other vegetation to worry about, so we could confine our considerations to structural matters. The gravel pit was at a grade about five feet lower than the undisturbed wooded landscape at its north edge.

So there wasn't much "site prep" at Earthwood, except for building up a pad of sand to float the slab on, described in the next chapter. Nevertheless, I still managed to make a mistake, which cost us additional earth-moving expenses later on.

As we were remote from any other house or septic systems and planned on remaining "off-the-grid" at Earthwood, we decided on a shallow well, as opposed to a deep drilled well requiring an electric submersible pump. We'd had the site "witched" by a neighbor whose success rate for finding water was, to all local knowledge, one hundred per cent. Ed Garrow would dig the well with a tracked excavator, sometimes called a "traxcavator." Think of a giant backhoe arm mounted on large tracks. The machine can excavate to 19 feet of depth.

The spot picked out for the well was at the edge of the gravel pit, but on the original grade, just 20 or 30 feet west of our intended house site. Ed set his machine on the gravel layer on the east side of the well, to gain a few feet of digging potential and proceeded to dig as far as he could, 19 feet. It was a bitterly cold December day. We looked down into a dry hole. I asked Ed what our options were and he told me that he could dig a hole, drive the machine into it, and dig the well deeper. I had no idea of the size of the hole which would result from this, and failed to communicate properly where we wanted the house. At 23 feet of depth, the traxcavator was maxed out again, but there was a little water in the hole, and we watched as the puddle grew bigger. We weren't deep enough for a well, so Ed "dug her in again," but this time, the hole needed

to be about four times larger – and the ramp twice as long – to gain an additional four feet. Now we were encroaching on area intended for the house. Had we dug the well from the south instead of the east, we would not have disturbed the earth where the house was meant to be.

We ended up with a great well – cold, clear, pure spring water for these last 24 years – but the house needed to be sited about 30 feet east of where it was supposed to be. This was not a disaster on our six acres of land, as it might have been on a half-acre lot. But our driveway is 30 feet longer than it needed to be (think snow removal) and we were not able to take full advantage of the five feet of extra grade, which meant more money spent on moving earth to create the berms after the house was built.

With the excavation done – at Earthwood, it was pretty much done when we bought the property – it was time to start bringing material in, and building the foundations, which is the thrust of the very next chapter.

Chapter 3

FOUNDATIONS

With all due respect to Ken Kern, I consider Henry David Thoreau to be the true father of the American owner-builder movement. In Chapter 1 of *Walden*, entitled "Economy," the philosopher admonishes us to build our castles in the air. "That is where they should be," says Henry (I feel that I know him on a first-name basis), who adds: "Then put foundations under them." While this is a nice piece of inspirational writing, it is actually easier to do foundations the other way around.

As with site prep, our experiences at Log End Cave and Earthwood were different. Let's do the Cave first.

LOG END CAVE: LAYING OUT THE FOOTINGS

A lot of building books wax eloquent on the benefits of setting up "batter boards" to find the four corners of the building. The idea of these grotesque, often flimsy homemade contraptions is that you can slide nylon line along them and check diagonals easily to assure that the

foundation is a true rectangle. Well, we battled with them for hours before a contractor friend arrived on site and told us they weren't worth the effort and that they would just be in the way of the backhoe. Right away we changed to the faster and tidier educated guesswork method.

Once the heavy equipment has got the site leveled with dimensions at least three feet greater than the outside dimensions of the footings, drive a wooden stake at – say – the northwest corner, two feet in from each side of the leveled area. Put a nail in the top of the stake, with its head sticking out an inch for tying a nylon mason's line. Buy a ball of this line at the building supply; you'll be using a lot of it. Measure for the outside length of the footing along the north wall to your determined point (35-feet-11-inches for Log End Cave), keeping about two feet in from the sloped edge of the excavation. Again, drive a stake in the ground, and set a nail into its top. Now comes the math part of the educated guessing, and I'll use the actual Cave plan numbers as an example. Our outside footing dimensions are 30-feet 8½-inches-by-35-feet 11-inches (30.71 × 35.92). Thanks to Mr. Pythagoras (remember $a^2 + b^2 = c^2$?), we can calculate for the hypotenuse (c):

$$c2 = a^2 + b^2$$
$$c = \sqrt{a^2 + b^2}$$
$$c = \sqrt{(30.71)^2 + (35.92)^2}$$
$$c = \sqrt{943.10 + 1290.25}$$
$$c = \sqrt{2233.35}$$
$$c = 47.258 \text{ feet} = 47 \text{ feet } 3 \text{ inches}$$

These calculations can be done quickly on a calculator with the square-root function.

Next, hook your 50-foot tape to the nail on the northwest corner, and, on the ground near the southwest corner, describe an arc with a radius equal to the west side footing dimension, 30-feet 8½-inches in our case. (The rather odd numbers come from our planned use of surface bonded blocks, explained in Chapter 5.) Now, hook the tape on the nail at the northeast corner and describe a second arc equal to the calculated diagonal measure, 47-feet 3-inches in our example. The point where the two arcs intersect is the southwest corner. Drive in a stake and set a nail there. Re-measure both ways and get that nail exactly on the right spot. The nails are the real points, not the stakes. Find the southeast corner in similar fashion. Check your work by checking the second (northwest to southeast) diagonal. The diagonal measurements must be the same for the rectangle to have four square corners.

Maybe the rectangle you've laid out doesn't use the space to best advantage. Maybe it crowds one slope, while there is more than enough room on the adjacent side. It is an easy matter to rotate the rectangle to alleviate the problem. You might even have to do a little shovel work if one of the sides is short of working space.

This calculated guessing more than one trial to get all four sides the right length, and the diagonals to check out. But it beats making batter boards that you're only likely to use once. In about twenty minutes, we had our four corners to within a half-inch, accurate enough for the footing.

For the benefit of the backhoe operator, place flags or white stakes on the various bankings, aligned with the outside edge of the footings. You can set these by eyeballing over the footing stakes, and it's even easier with two people. In all, you'll place eight such guide markers.

The Footing Track

Please look at Figure 3.1. To work out the footing track depth into the undisturbed earth, the important relationship to keep in mind is the one between the top surface of the footing and the top surface of the floor. Cross-hatching represents undisturbed earth. For underfloor drainage, allow for at least 4 inches of compacted sand to lay the floor upon.

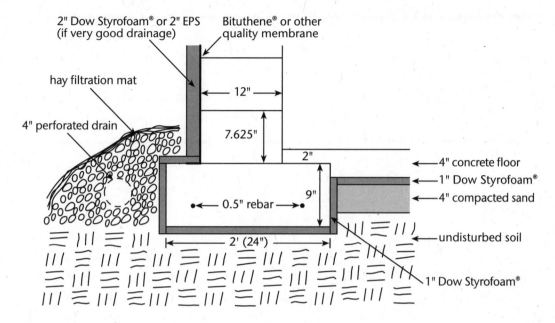

2" Dow Styrofoam® or 2" EPS
(if very good drainage)

Bituthene® or other
quality membrane

hay filtration mat

4" perforated drain

12"

7.625"

2"

0.5" rebar

9"

2' (24")

4" concrete floor

1" Dow Styrofoam®

4" compacted sand

undisturbed soil

1" Dow Styrofoam®

Fig. 3.1:

Footing details

for Log End

Cave, showing

correct

insulation

technique.

Our plan was for a four-inch-thick concrete floor. In reality, if you've got an honest three inches at the floor's thinnest portions, you've got a strong floor, but four inches is nice for extra thermal mass. With in-slab heating tubes, seen in the next chapter, a six-inch-thick floor is often specified for even greater mass. Note that our plan calls for an inch of extruded polystyrene, such as Dow Blueboard™, under the floor and all around the footings, and this comes into the footing depth calculations, too. With in-slab radiant floor heating, two inches of extruded polystyrene is specified.

The detailing in Figure 3.1, based on nine-inch-thick concrete footings, shows that only three inches of undisturbed earth needs to be removed to set up the footing forms and place the rigid foam insulation. In the actual event, we went 12 inches deep with the footings. A year later, at a workshop we co-instructed, concrete underground house expert Paul Isaacson pointed out to me that a 12-inch-thick footing is overkill. "Beyond nine inches, you're really wasting money on concrete, Rob," Paul told me. "We're not building a skyscraper here."

So the drawing shows a three-inch-deep track into undisturbed earth, whereas we actually went down six or seven inches, which is why we had the track done with a backhoe. Also, we knew that there were a lot of heavy stones lurking in the earth on this site. If you do remove a large stone from the footing track, be sure to compact any replacement earth to prevent differential settling in the future.

The other improvement in the drawing, which we did not do in the actual event, is very important, so I'll give it its very own heading.

LOG END CAVE: INSULATING THE FOOTINGS

We did not insulate around the footings at Log End Cave and our failure to do so caused condensation problems at the base of the wall during warm moist conditions in May, June, and July, a time when the footings were still conducting "coolth" from the still-cold soils at seven feet of depth. It wasn't until the footings warmed up near the end of July that the condensation disappeared. Wrapping the footings (and the floor) with Dow Blueboard™ prevents this condensation by keeping the concrete temperature up above dew point, as we have seen at Earthwood. Figure 1.2 illustrates the point.

Dow Styrofoam® Roofmate® and Styrofoam® High Load 40 Blueboard™ have compression strengths of 40 pounds per square inch (5,600 pounds per square foot) with only 10 percent deflection, and Dow makes others with even higher compression strength. These Dow products are the only rigid foam insulations that I'm aware of with sufficient strength to be used under the footings. (Other extruded polystyrenes would be okay at other locations.)

As the load on houses like Log End Cave is around 1,800 pounds per square foot, deflection (compression) will be considerably less than 10 percent. Don't use any of the expanded polystyrenes ("beadboard") around the footings or under the floor, as they are not water resistant.

The footing track needs to be square, level and about 30 inches wide, in order to accommodate the 24-inch-wide footing, two forming boards, and a couple of one-inch pieces of rigid foam insulation.

RESISTING LATERAL LOAD ON THE WALL

Figure 3.1 shows the method we used to resist lateral pressure at the base of the wall. The top two inches of the 4-inch floor slab (which is poured on a different day) acts as a foot against the first course of blocks moving inward. The order of events is to pour the footings one day, mortar the first course of blocks on a different day, and pour the floor on yet another day. All this is covered in the next chapter about the floor pour. The method worked, but was cumbersome. The use of a keyed joint is a proven alternative that might be a little easier to do. I'd try it next time.

The keyed joint shown in Figure 3.2 can be cast into the footing by setting a piece of wood flush with the surface of the concrete when the footing is poured. You can make a good keyway form by rip-sawing a regular 2-by-4 down the center with a circular saw. Set the saw's blade angle at about 75 or 80 degrees instead of 90 degrees to create a draft angle in the wooden keyway form, thus facilitating its easy removal later on. Other ways to make the wooden form easy to remove later are: (1) oil the board before setting it in the concrete and (2) after the concrete sets, install stout deck screws into the key piece, left sticking out of it to provide something to grab. Note in Figure 3.2 that the draft angle is kept to the outside of the footing. After the concrete has set, remove the wooden forming piece.

Later on, the first course of blocks can be firmly tied to the footing by filling the block cores halfway with concrete, which finds its way into the keyway you've cast, thus locking the blocks to the footing. The keyway would also be valuable if you decide to pour the walls instead of using blocks.

I like this method because the concrete floor can be poured before the first course of blocks (or poured wall) is installed and nothing is in the way of screeding and finishing the floor. If you go with a six-inch-thick floor — say, for radiant in-floor heating — or go with a thicker bed of compacted sand under the floor, you really don't need to set the footing tracks beneath original grade at all. On the other hand, setting the tracks down even three inches will help to prevent the forming boards from "blowing out" during the actual pour.

THE FROST WALL

In northern climes, it is necessary to protect footings close to the surface from the uplifting pressure of frost heaving. Heaving occurs when wet ground beneath the footing freezes and expands. The massive weight of the wall is of little help in resisting frost expansion pressures. The only solution is to make sure the ground beneath the footings doesn't freeze. Clearly, that won't happen

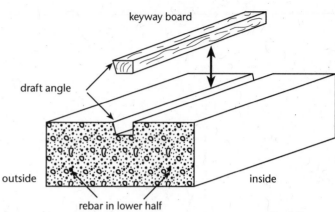

keyway board

draft angle

outside

inside

rebar in lower half

Fig. 3.2:
A keyed joint can be created when an oiled keyway board is removed from the concrete after it is set. The draft angle should be kept to the outside of the footing.

on the three sides of Log End Cave which are six feet or more below grade, but part of the south wall is a problem, and the entire south wall footing would need protection in the case of a house like Richard Guay's (Figures 2.2 and 2.3) Protection against freezing can be accomplished by going deeper with the footing (down below local frost depth) or protecting it with extruded polystyrene, or both. Check with your local building department for the working frost depth where you live.

At the original Log End Cave, there is about eight feet of wall near to and including the doorway, the eastern quarter of the south wall. The footing below this section of the wall doesn't have the protection of the three feet of earth that we used to berm up to the underside of the three large south-facing windows. Our solution was to increase the footing depth to a full 24 inches over a 12-foot section of the southeast corner. This extra concrete has worked fine. In 28 years, there has been no heaving of any kind on the structure. In theory, the frost wall should have been 48

*Fig. 3.3:
Frost
penetration
follows a 45-
degree angle
from the
surface, so
footings can
gain protection
by the
placement of
extruded
polystyrene
laterally away
from the
building, as
shown. Insets
show alternative
insulation
detailing.*

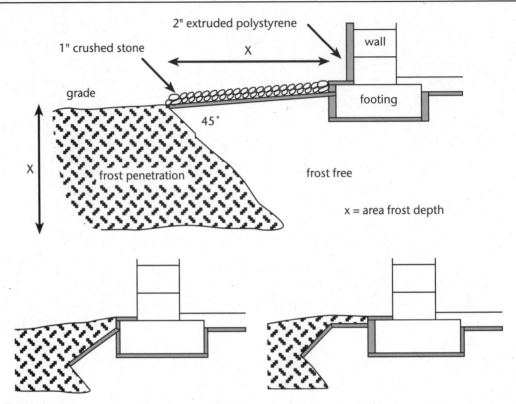

inches, the working depth for our area. But this protected south-facing wall – a suntrap, really – doesn't experience that degree of frost penetration. Extra care would be imperative for unprotected walls on other sides of the building.

Another method of protecting against frost heaving has gained favor since the advent of closed cell insulation. The method takes advantage of frost's characteristic of permeating into the soil from above at a 45-degree angle, as seen in Figure 3.3. The advantages of this system are savings on materials (much less concrete) and, in many cases, easier construction.

Frost protection comes from extending a minimum of two inches of closed-cell extruded polystyrene away from the foundation. The width of this insulation curtain should be the same as the engineered frost depth for your area, four feet, for example, in the northern tier of New York State. There are various ways the insulation can be configured including the two smaller insets at the bottom of Figure 3.3. For thorough examinations of this method of insulation, see two reference manuals listed in the Bibliography under "Frost-Protected Foundations." Both of these manuals can be viewed or downloaded free from the internet.

Protection from frost-heaving is a structural consideration which must be addressed. But the prevention of energy nosebleed into cold (but not frozen) soils is also important to prevent condensation around the interior perimeter of the home. This is important all around the home's perimeter, both below-grade portions and at any frostwall areas. Lateral insulation out from the edge of the footing, or – in the case of Earthwood – the floating slab, will protect against frost damage, but will not adequately protect against energy nosebleed and the condensation that follows. This is why Figure 3.3 calls for extruded polystyrene under the footings and floor as well, particularly in cold climates. Under the footings, use Dow Blueboard™ for adequate compression strength.

Your choice for frostwall protection will be a function of cost and convenience.

EARTHWOOD: THE FLOATING SLAB

Also known as a *slab-on-grade* or, sometimes, an *Alaskan slab*, the floating slab was one of Frank Lloyd Wright's favorite foundation methods because it attends to the problem of potential damage from frost heaving in the most basic of ways: it drains water away from under the foundation. No water, no freezing. No freezing, no heaving. We used the floating slab at the main house at Earthwood built in 1981, and at eight additional outbuildings on site since then. The outbuildings are not earth-sheltered, but all except one – our garage – have heavy living roofs. Needless to say, we have become enamored of the technique because of its relatively low cost, ease of construction, and successful performance. Here's how to do it:

First, all organic material like sod, grass, and topsoil needs to be scraped off the site and piled off-site for use, later, on the roof. You want to get down to "undisturbed earth," typically six to 12 inches below grade, although this can vary. Next, we build up a "pad" on top of the undisturbed earth, using a good percolating material such as coarse sand, sandy gravel, or ½-inch (#1) or 1-inch (#2) crushed stone. Whatever material you use, compact it in layers – called *runs* – of about six inches of thickness at a time. Sand or sandy gravel needs to be compacted wet with a powered compactor hired from a tool rental store. Pounding dry sand is an exercise in futility.

A sand pad. We built the pad at Earthwood out of the indigenous materials of our gravel pit, specifically a vein of sand about 40 yards south of the house site. Ed Garrow, our operator, showed us a bulldozing trick. He pushed the sand ahead of his bulldozer blade towards the site. The sand, of course, spilled out each side of the blade before the dozer was halfway there. Ed seemed unperturbed and backed up for another push. I knew Ed well enough by this time to bite my tongue. The next large push also spilled out, making larger berms on each side of Ed's path. After a few pushes, Ed was guiding full loads of sand to the pad site every time, the berms creating a kind of trough to guide the material to its destination. To complete the job and tidy things

*Fig. 3.4:
Dennis Lee
tamps the sand
pad at
Earthwood with
a power
compactor.*

*Fig. 3.5:
Three or four
runs of sand are
built up and
individually
compacted,
resulting in a
final compacted
pad of at least
18".*

*Fig. 3.6:
A view of the
finished sand
pad at
Earthwood. The
pad extension
near the truck is
for the
greenhouse.*

diameter, we created a pad 46 feet in diameter. A 30-foot-by-40-foot slab, for example, would float on a pad measuring 36 feet by 46 feet. We set two-by-four depth stakes all around the pad location, with red surveying bands tied around the stakes at final pad grade level, about 16 inches above the undisturbed earth. Remember that at Earthwood's site in a gravel pit, "undisturbed earth" was the pit's flat surface itself, compacted by more than 20 years of rain and snow load.

I hired Dennis Lee for five months to work on the project. After Ed would build up a run of six inches of sand, Dennis and I would water and compact the material while Ed gathered more sand at the other end of his trench, or spread gravel on the new driveway. We would compact in both primary compass directions (north-south, east-west) and then a third time in a spiral configuration.

Every square foot was hit three times with the compactor. Then we'd call Ed for more sand for the next six-inch run.

By the end of the second day, we had a well-compacted flat pad about 18 inches in depth, seen in Figure 3.6.

If good percolating material is available on site, use it for your pad. On sites with good drainage, frost heaving is unlikely anyway. Most people will need to bring material in, however, so you will need to calculate the volume of sand or gravel you need, adding about 20 percent to your volume figures for compaction.

Gravel – sometimes confused with crushed stone – is a mix of sand and small stones, and

up, Ed made the final passes by angling his blade, pushing one berm's material to site, supported by the other berm. He smoothened the path with material from the second berm.

The pad should be a full six feet wider in each dimension, to accommodate for a sloped skirt around the slab's edge. As Earthwood was round, with the outside of the footings almost 40 feet in

should be treated in the same way as sand. Larger stones can be handpicked from the pad during construction.

A crushed stone pad. Crushed stone provides excellent drainage. It simply doesn't hold water, providing you don't place it in a dish or lens of impervious material. The disadvantage with crushed stone is that it is hard to dig into for the setting of footing forms. One way to do this is to set the footing forms on the second or third run of crushed stone, and then to spread the last run of stone carefully by hand.

Another problem with crushed stone is that it is a more difficult material in which to lay under-floor plumbing, discussed in the next chapter.

Opinions vary on whether or not crushed stone needs to be compacted. Some say – I *used* to say – that crushed stone is compacted when it falls off the truck. But I have changed my view on this. Railroad tracks float on a bed of compacted crushed stone. And I once saw a stackwall-cornered cordwood home floating on a railroad tie foundation (on crushed stone) where the crushed stone was not compacted. One of the corners had collapsed slightly downward because of differential settling. Compacting the gravel bed would have prevented this.

The "Monolithic" Floating Slab

You have two choices with a floating slab: pour the footing one day and the floor on a different day (as we did at both Log End Cave and Earthwood) or pour the footings and floor on the same day (as we have done with most of our outbuildings), forming what's called a monolithic slab. Literally, monolithic means "of a single stone." If you are not experienced at concrete work, or with a house-sized project, I strongly advise the two-step approach. There are different sorts of preparation for the footings and the floor, so it is less mentally taxing to handle just one process at a time. It is also much less physically taxing, because you'll only have about half as much concrete to pour each day if you separate the two jobs. Finally, there is a great advantage in being able to use the footings as a guide for screeding a nice level concrete floor.

With small buildings, such as our new octagonal guesthouse at Earthwood, it is an easy matter to screed from the footing forms to a block of wood set in the middle of the building. Also, the underfloor plumbing was very simple at this guesthouse. We had a good crew of six bodies to help, and only four yards of concrete to pour, so a monolithic slab made sense. See Figure 3.7.

If you are having the slab done by professionals, they will probably be happier doing any project monolithically. It is less work for them to come just one day instead of two, and they are used to laying concrete with or without benefit of footings to screed to. Just make sure that they, or you, have attended to all of the preparations for the footings and floor which follow in the conclusion of this chapter, and the next.

Fig. 3.7: A crew of six made easy work of pouring the small four-cubic-yard monolithic floating slab at the new octagonal guesthouse at Earthwood.

EARTHWOOD: LAYING OUT AND DIGGING THE FOOTING TRACKS

Some people who are comfortable with laying out and building rectilinear structures cringe at the thought of a circular home, and yet the circle could be argued to be much easier to lay out … and build. So-called "primitive" people, as well as the other building species, all build round. (Maybe because it *is* easier!)

It's also a delight to do layout on the compacted sand pad, already described above. It is easy to draw on the sand with a stick, and it is easy to excavate footing tracks.

Everyone knows that a circle has just one unique point: its center. And most people readily understand that it easy to draw a circle by drawing a continuous arc based upon the center. On a piece of paper, we use a drawing compass. The pointy end of one compass arm pricks the center

point of the building, the arm with the pen or pencil swings around the center, creating a perfect circle. On the sand pad, we do the same. A nail sticking out of a stake in the middle of the site marks the exact center of the building. Clip a tape measure on this nail and stretch the tape out to the correct radius – say 20 feet – and walk around the site while inscribing a visible groove in the sand, using a stick set at the 20-foot mark. *Voila,* a perfect circle!

Now I will complicate this simple discussion, but only a little. We wanted to take advantage of solar gain in the house, so the north-south meridian was important. I also wanted to be able to use the home's 32 radial rafters like the points of a ship's compass card, so that I could find any celestial object by going to the appropriate rafter. None of this may be important to the reader, but what should be important is having a number of external reference points to use for finding important features in the home, such as internal post locations, the toilet receptacle, etc. For us, with our internal octagon framework, eight equidistant external points proved to be very useful. Combining the need for these reference points with my astronomical predilections, I decided to place directional stakes at the eight primary compass points: north, northeast, east, southeast, etc. To get the required accuracy of orientation, I called on my good friend George Barber.

We enjoyed a fascinating evening of beer and blether, during which George described several wonderful means of finding true north, some making use of the stars and sun. (I have already described how to use the time of sunrise and sunset to do this: halfway between those times, the sun is in solar south.) Finally, George asked me: "How accurate do you want to be?"

"Can you get us to within a half of a degree?" I asked.

George smiled and said, "I can get you to within a half a minute."

The next day, George brought his wonderful transit to the site, and, using a very accurate magnetic declination adjustment, we laid out the eight accurate compass points, composed of a

stake and a nail, each about a yard outside the footing tracks, and each equidistant from the center and from each other. George, a professional surveyor, did this all with a transit and all nails were equidistant to an accuracy of 1/16-inch.

If you don't have a George Barber, follow the steps of Figure 3.8, which show you how to accomplish the same goal with a tape measure and a sharp stick for drawing in the sand.

LOG END CAVE: FORMING THE FOOTINGS

With a rectilinear construct, like Log End Cave and most American buildings, form the footings with two-inch-thick planks, such as a 2-by-8 or 2-by-10. At the Cave, we used 2-by-10 forming boards lent to us by a contractor friend. These worked out well with our 12-inch-deep footings; we just let the extra two inches of concrete take the shape of the slightly deepened track. But, as stated earlier, I now advise nine inches of concrete depth for the footings, and an inch of Dow Styrofoam® Blueboard™ under the footings in northern climates. Again, this would have worked perfectly with two-by-ten forming boards.

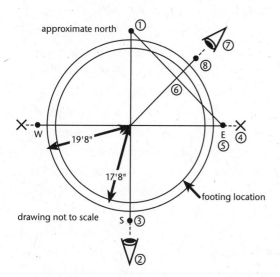

Fig. 3.8: 1. Drive north stake 22' from center. 2. Extend the line from the north stake through the center stake and on in a southerly direction. This line can be eyeballed quite accurately. 3. Drive south stake 22' from center. 4. To find the east and west points: Clip one end of a 50' tape to the south point nail, and describe an arc in the sand near the eastern (and western) locations, using a radius of about 1.5 times the center to south distance, or 33' in this example. Now clip the tape to the north nail and draw two more 33' arcs in the sand, intersecting those already drawn. The intersections are exactly east and west of the center. 5. Clip your tape to the center nail and have someone hold the tape so that it passes exactly over the east intersection. Drive a stake 22' from center and mark the east point with a nail. Do the same for west. 6. To find northeast: Measure the distance from the north to the east points. Mark the ground accurately at a point exactly halfway between. 7. Extend your tape from the center point through the ground mark and straight on beyond. 8. At 22', drive the northeast stake. Similarly, use steps 6 thru 8 to find southeast, southwest and northwest.

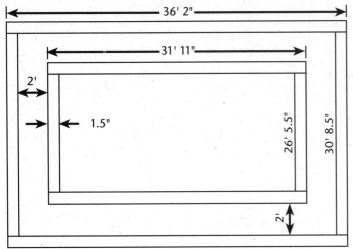

outside footing dimensions: 35' 11" × 30' 8.5"
inside footing dimensions: 31' 11" × 26' 8.5"
drawing not to scale

Fig. 3.9:

This diagram was used to work out the lengths of all the forming boards at Log End Cave.

It is worth taking the time to figure out your footing form sizes and cut them to their exact required length to place in the slightly oversized footing tracks you have already sculpted into the undisturbed earth (or pad). The only way I can do something like this is to make a drawing with all the pertinent dimensions shown. Figure 3.9 is a slightly tidied-up copy of the drawing I used to help me make the Cave footing forms. The outside footing dimensions needed to be 35-feet 11-inches (east-west) by 30-feet 8½-inches (north-south). Notice, then, that the long outer forms are made three inches longer, because the shorter forms (cut to the actual north-south footing dimension of 30-feet 8½-inches) will eat up three inches of the east-west dimension because they are set inside of the outside east-west forming. (The

three inches is twice the 1.5-inch thickness of finished "two-by" lumber.) Similarly, the east and west planks of the inner set of forms are three inches short of the inner measurement of the footing because the thickness of the planks running the other way make up the difference.

You should be able to determine where all the numbers are coming from by studying Figure 3.9. Hopefully, your numbers will be easier. The odd numbers we used are a function of using funny numbers that are a part of surface-bonded block work, as you'll see in Chapter 5. With poured walls or mortared blocks, you won't be dealing with obtuse fractions.

Forming boards can be cleated together with three-foot sections of planking left to the outside of the footing track, as per Figure 3-10. We actual used the little two-inch spacer shown in the diagram on the long (36-foot 2-inch) sides, so that we could make full use of my friend's 18-foot forming boards. Normally, you'd butt plank to plank. In the old days, we used double-headed forming nails, also called duplex or scaffolding nails. Nowadays, you have the option of screwing your forms together, but clean the wet concrete off the screw heads for easy removal later on. If you borrow forming boards, be sure to return them just as clean as you got them.

You'll need enough 24-inch 2-by-4 stakes to place one every four or five feet all around the outer sides of both the interior and exterior sets of footing forms. Allow a few extra for the ones you destroy with the sledge hammer, and hit the stakes, not the forms. Fifteen eight-foot 2-by-4s will make sixty 24-inch stakes. Put tapered points on each one.

Place the footing forms so that their tops are all at the same level. This job is difficult without a contractor's level. Another great modern tool you can hire is a laser level, which spins a strong beam of light around the site a convenient height off the ground. In Figure 3.11, a very young Jaki holds a grade stick while a friend reads the number … and I take the picture.

For your first practice with the contractor's (or laser) level, determine the average grade of the bottom of your footing tracks. Take three or four readings at equidistant points along each side of the track – or a dozen readings around a round track – and average the readings with a calculator. The average grade will be useful in setting the footing forms sensibly, but this exercise might also reveal some egregious discrepancies in the depth of your track, which you will want to correct immediately with hand tools before proceeding any further.

Using the existing corner stakes as a guide, bring in the longer of the north side forms (for example) and put it roughly in place. A few inches in from each end of the long form board, drive new stakes into the ground, always on the outside of the forming board, never where concrete is poured. Drive all new stakes so that they are at the same grade, as judged by the contractor's level or laser level. This will be about ten inches above the average grade of the depth of the footing tracks,

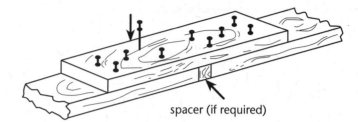

spacer (if required)

Fig. 3.10:
Planks can be cleated together as shown, using forming nails or screws.

depending on the details you have chosen in your footing cross-sectional diagram, which will look something like Figure 3-1. Your footing cross-section diagram should include all the details of your floor, such as footing dimensions, floor thickness, insulation, and your keying method.

Nail or screw the long north side form to the new stakes so that the top of the form is level with

Fig. 3.11: Jaki holds a grade stake while a friend uses a contractor's level to assure that the inner and outer footing forms are all at the same grade.

the top of the stake. Eyeball the form straight and drive a third stake into the ground about halfway along its length. Only the corner stakes need to be at the same grade as the top of the forms, so that they can serve as benchmarks. Other stakes can be driven just slightly lower than the tops of the forms, so that they will not get in the way of screeding the concrete as it is poured.

Next, level the form. This is most easily done with three people: one to hold the grade stick, one to read the contractor's level, and one to pound the nails or install screws from the stake to the forming boards. The beauty of the laser level is that once it is set up and spinning, you don't need a person to read the level. The laser light creates a perfect horizontal plane as far as the light can reach, and it shows up very clearly on your grade stake.

Put at least one stake between the corners and the midpoint of each individual forming board, two or more on exceptionally long spans. Check the entire top surface for level.

The other seven form boards are done as described above. Complete the outer ring of forms before proceeding to the inner ring. And recheck the diagonals with the 50-footer to make sure the forms are square. The inner ring of forming boards is placed to leave a space equal to the width of your footings, which was two feet at both Log End Cave and at Earthwood. Figure 3.9 shows the tracks as two feet wide. But, don't forget the inch of insulation on both sides of the footing, as per Figure 3.1. You can either put those two pieces of one-inch insulation up against the inner sides of the footing forms (which yields a truly 22-inch-wide concrete footing, still substantial), or you can add the two inches for insulation to the dimensions of your footing tracks and make the corresponding correction to your forming diagram, the one resembling Figure 3.1.

Check the tops of both the inner and outer sets of forms, to make sure they are all at the same elevation, and, therefore, level with each other. Use a lever, such as a 36-inch J-bar, to move stakes upward slightly, a sledge to set them in a wee bit. Always hit the stakes, not the forms themselves. It may be necessary to remove a little earth beneath the forms, in spots, so that they can be taken to their proper grade.

The heavy wet concrete wants to spread the forms, and the stakes may not be enough to resist this bowing (which, in the extreme, can become "concrete blowout," which tends to spoil the day.) There are various ways to resist this bowing. You can jam 2-by-4-type material between the forms and the excavation banking. You can throw loose earth up against the base of the forms and pack it with a sledgehammer. (This method also prevents leakage of concrete out of the bottom of the forms.) And you can make a few moveable cleats like the one shown in Figure 3.12, and have them at the ready for use wherever needed.

With the forms in place and well braced, the insulation can be installed at the bottom of the footing tracks, and on its inner sidewalls. Another advantage of using rigid foam insulation here is that you will not have to oil the footing forms to facilitate their removal after the concrete sets. (Paint old engine oil onto the forms wherever concrete is to be poured directly against wooden forms, as may be the case in the south or as we did on the top four inches of the inner form at Earthwood.)

Please see the Sidebar on page 66 for my commentary on rigid-foam insulation choices.

Reinforcing Bar (Rebar)

Reinforce the footing with a *minimum* of two pieces of ½-inch (#4) reinforcing bar (also known as "rebar", which shortened term I will use hereafter). Building code in your area may require three pieces, and might insist upon ⅝-inch (#5) rebar, but we have had no problem at Log End Cave and Earthwood with two pieces. The rebar runs laterally around the footing track, and should be placed in the bottom half of the footing pour, as seen in Figure 3.1, to give the footings the tensile strength against settling. This is the primary purpose of rebar. Keep the steel bars a minimum of three inches in from the bottom and edges of the track, by supporting it during the pour on clean pieces of broken bricks or three-inch-thick washed flat stones. Special wire supports made for the purpose are available, called *frogs* or *chairs*, but I have never used them. Whatever supports you decide on, they'll become a part of the concrete, so choose something clean, strong and the right thickness, usually three inches. Wood supports are unacceptable.

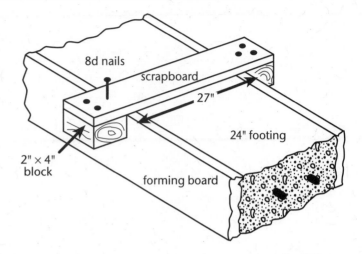

Fig. 3.12: Moveable cleats like this one will help to keep the footing forms from spreading under the pressure of the concrete.

Rebar comes in 10-foot and 20-foot lengths. Fewer overlapping joins are required with the 20-foot pieces, but they are harder to carry home from the supply yard. I usually put a soft protection blanket on the roof of my little pickup truck and bundle and tie the pieces together to the front and back bumpers. Sometimes, as with the octagon we recently poured, and with other small buildings, the 10-foot pieces have been very convenient. The rule for joining any rebar in a track is to overlap pieces by forty times (40×) the diameter of the rebar. So, with ½-inch rebar, the overlap is 20 inches. With ⅝-inch rebar, sometimes specified, the lap would be 25 inches (⅝" × 40 = 25 inches). Tie adjacent rebars together with twisted forming wire made for the purpose. Buy a small reel of this wire for the project. It's handy stuff. After 20 years, I finally used up the last of my reel, and now I miss it.

On Below-grade Insulation

There are a variety of rigid foam insulations available, including polyurethane foam, polystyrene, and others. Conventional wisdom, backed by studies done during the heyday of earth-sheltered housing, points towards "closed-cell" extruded polystyrene as the safest bet below grade. Polyurethane foam, with its excellent dry R-value of R-8 per inch, can absorb moisture in unprotected applications, causing a great reduction in its R-value. Having said that, a friend who scored a good deal on a large quantity of aluminum-foil faced polyurethane foam, used it on the roof of his one-story Earthwood house, beneath his living roof. After several years, I have not been able to find any deterioration in the two-inch-thick sheets he used. The house is still very energy-efficient. I expect that the aluminum surface helps, but his roof also drains well.

Conventional wisdom also steers builders away from expanded polystyrene (also know as EPS, or "beadboard"). As EPS is not a closed-cell structure, moisture can more readily transfer through the material, turning it into a sodden sponge. I have seen this happen to cheap beadboard buried just below the surface of poorly drained soils. On the other hand, we used EPS on the sidewalls of Log End Cave, and, a few years after construction, I had occasion to dig down a sidewall about six feet to install a new waterline through the wall. The EPS was just as dry and crisp as the day it was installed. I credit its performance to our use of good percolating sand backfill which takes groundwater down to the footing drains.

I have had good luck with extruded polystyrene below grade, but, even here, I have seen the so-called closed-cell Styrofoam® get heavy with water in poorly drained below-

We actually used ⅝-inch iron silo hoops as rebar at both Log End Cave and Earthwood. We had to do a certain amount of hacksawing to salvage them, but, in those days time and money were always being weighed up, one against the other. In point of fact, construction rebar is not very expensive.

EARTHWOOD: A ROUND FOOTING

The placement of Earthwood's footings – nine inches deep by 24 inches wide – was a function of the round cordwood and concrete-block walls they will support. The walls are 16 inches thick and need to be centered on the footings for bearing and stability. As the home has an inner

grade locations. Where have I arrived at from all these mixed signals? For me, the answer is to err on the side of caution. I will continue to use extruded polystyrenes such as Styrofoam® and Owens-Corning pink Foamular™, but it is really important to use good drainage against and around the material. Protection of the foam from sodden soils (particularly on a gently pitched roof) can be greatly improved by the use of the layer of 6-mil polyethylene described in Chapter 7 and Chapter 8 about living roofs.

Finally, on the bottom of the footing track, use Styrofoam® Roofmate® or Styrofoam® High Load 40 Blueboard™ for its superior compression strength, just 10 percent deflection under a load of 5,600 pounds per square foot. I have never found Styrofoam® to be any more expensive than other extruded polystyrenes. Do a local cost comparison yourself. You might want to stick with the Blueboard™ for all below-grade insulation.

I am sorry to report that I do not know of any "natural" products that will stand up to conditions below grade. On the positive side, extruded polystyrene manufacturers have removed the offensive CFCs (chlorofluorocarbons) from the manufacturing process, those same CFCs famous for their contribution to the depletion on the earth's all-important ozone layer. Dow Styrofoam®'s website, listed in Appendix B, says: "Today, all 25 Dow plants worldwide manufacture Styrofoam Brand products with either HCFC blowing agents, which have an ozone depletion potential less than 10 percent of standard CFC blowing agents, or even HCFC-free blowing agents as a permanent, ozone-friendly foam."

In the long term, these insulation products, in my view, justify their existence because of the tremendous amount of energy they save. In well-drained earth-sheltered housing, they will last indefinitely. That is, they are not biodegradable. At first blush, that sounds like a bad thing, but, if you think about it, you don't want degradable insulation next to your home.

radius of 18 feet exactly, the inner radius of the footing forms needed to be 17-foot 8-inches. As the footings are two feet wide, the outer radius of the form boards is 19-foot 8-inches, as seen in Figure 3.8. The maximum saturated load of the home – two full stories plus an earth roof – when distributed on the two-foot width of the footings, comes out to about 1,800 pounds per square foot (psf.) I spent time designing the building in such a way that all load-bearing components of the foundation – footings, pillar footings, and the central mass (masonry stove) footings would all support similar kinds of numbers in the range of 1,600 to 2,000 psf, preventing differential settling.

Fig. 3.13:
Left: Footings
formed and
poured
independently
of the floor slab.
Right: Footings
and floor
poured at the
same time, the
"monolithic"
slab.

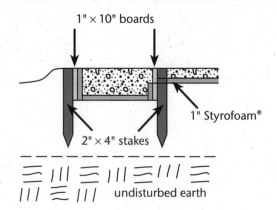

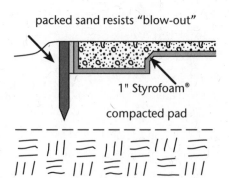

If some parts of the foundation support 3,000 psf and other parts just 600 psf, well, these are dissimilar numbers, and could result in problems as the different parts of the building settle into the sand pad – or even undisturbed earth – to different degrees.

We actually used 1-by-10-inch boards to form our nine-inch-deep by 24-inch-wide footings – there would be an inch of Dow Blueboard™ at the bottom of the track – but we had to keep planing the boards thinner and thinner until, finally, at ⅝-inch thick they would bend around the curve. Never again. I soon learned that the easiest and cheapest forming material for round footings or slabs is ¼-inch plywood. A full-sized 4-by-8-foot plywood sheet can be ripped into five "boards" at 9½ inches wide, so eight sheets will yield 320 lineal feet of forming material, plenty for our project. Quarter-inch plywood is very strong and bends nicely even on tight circles, as we found when we used it for the seven-foot-diameter central mass footing, discussed below.

You do need to stake it quite frequently, however, to prevent it from bowing out during the pour, or worse, blowing out.

The left-hand side of Figure 3.13, above, shows the footing tracks floating on the sand pad at Earthwood. The right-hand side of Figure 3.13 shows the detail if the floor and footings had been poured at the same time, the monolithic floating slab shown in Figure 3.7.

Inner and outer forms are set at the right grade and leveled, as already discussed in the Cave example. Instead of being concerned about square, though, the salient feature of a round building is that all points of any ring be equidistant from the center.

The only other tricky bit with the round footing track is installing the horizontal insulation below the pour. I discovered that a 2-by-8-foot sheet of one-inch Dow Blueboard™ (extruded polystyrene with high compression strength) could be cut to fit almost perfectly between our footing tracks, with just three cuts, as seen in

Figure 3.14. Even with a different diameter, or different footing dimensions, take a little time to work out the best use of a 2-by-8-foot (or 4-by-8-foot) sheet of insulation. With two-foot wide tracks, a good choice for the sizes of building we're talking about, a slight adjustment in the numbers in Figure 3.14 will allow the pieces to fit neatly with any diameter from 30 feet on up.

We reinforced our footings with two rings of #5 (⅝-inch) rebar, supported on three-inch-thick pieces of clean broken bricks, as seen in Figure 3.15. Keep all rebar in the lower half of the pour, where it adds tensile strength to the footing, and at least three inches from the edge of the pour.

The Buttresses at Earthwood

The cylindrical Earthwood home – both one- and two-story versions – has been built many times all over North America. If one side of the cylinder is to be loaded with a heavy berm, then there needs to be some kind of resistance incorporated into the foundation to prevent the home from moving with that load. At Earthwood, the earth berm on the northern hemisphere is as high as 13 feet and weighs an estimated 500 tons.

Think of Earthwood as a stone or brick arch, lying on its side. Arches get their amazing strength from the fact that masonry units, such as bricks, blocks, and stones, are extremely strong on compression. In his *Structures: Why Things Don't Fall Down*, English engineer J.E. Gordon – my favorite structural writer – says: "Elementary arithmetic shows that a tower with parallel walls

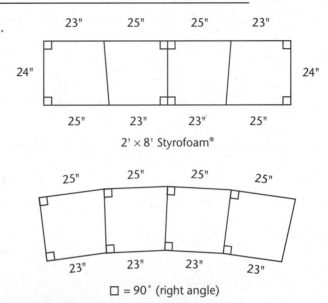

2' × 8' Styrofoam®

□ = 90˚ (right angle)

Fig. 3.14:
Two-foot by eight-foot sheets of extruded polystyrene, cut as shown at the top and installed as seen at the bottom, fit the footing tracks at Earthwood almost perfectly.

Fig. 3.15:
The footings are ready to pour.

could have been built to a height of 7,000 feet or 2 kilometres before the bricks at the bottom would be crushed." Figure 3.16 shows a typical arch. The vertical loads on the arch are transferred

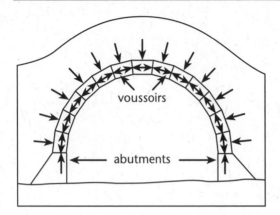

to lateral loads by the tapered masonry units, called the *voussoirs*. The abutments of the arch — the essential supports at each end of the span — must be strong enough to provide a reactionary thrust against the accumulated load of the voussoirs. The underside of the arch, you see, is on compression, and we do not begin to approach compression failure of the masonry units.

It was another English engineer, my friend John Jacobson, who pointed to the need for a similar kind of reactionary load or thrust at Earthwood, which is like an arch lying down, as seen in Figure 3.17. The voussoirs, in this case, would be eight-inch-wide corner blocks, laid transversely in the wall like log-ends.

But what would I use for abutments? After discussions with John, I designed the buttress system illustrated in Figures 3.18, 3.19 and 3.20.

The strength of the buttresses is predicated upon the reinforcing bar inside of it. I had all the necessary parts (shown in Figure 3.20) made at our local steel service center, all made from heavy ¾-inch (#6) rebar. The two square-shaped pieces were heat-bent and welded to close the square. The two diagonal pieces are bent at a 45-degree angle as shown. All parts of the buttress cage are tied to the footing reinforcing bar with forming

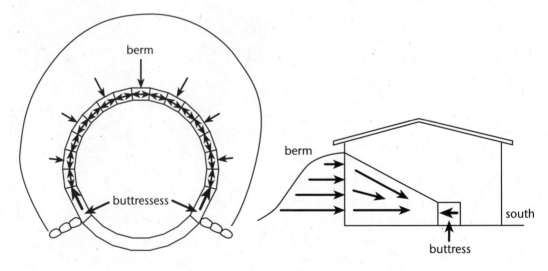

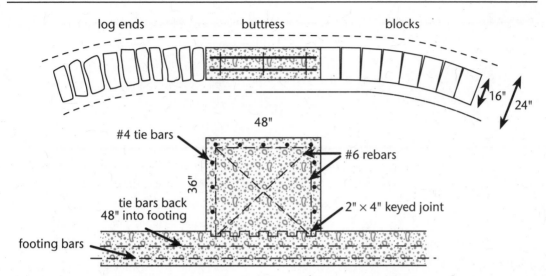

log ends   buttress   blocks

16" 24"

48"

#4 tie bars   #6 rebars

36"

tie bars back
48" into footing

2" × 4" keyed joint

footing bars

Fig. 3.18:
Buttress detail.

wire. Architect Marc Camens suggests using two-by-four pieces to create transverse keyed joints as shown in Figure 3.18.

Although we didn't do it, I think it is a good idea. Oil the little forming blocks so that, later, they can be easily removed from the concrete after a few hours. After the footings were poured, we built a forming box around the buttress cage assembly sticking out of the footings, which will be seen in the next chapter about floors.

The buttress cage as described was placed within a 12-inch-wide buttress form, described later, because we planned to do a four-inch cordwood facing on the interior to hide it from view. If I were doing it again, I would make the buttresses 16 inches wide, the same as both the cordwood wall on the southern hemisphere and the 16-inch block wall on the northern hemisphere.

Stone Mass Footing

The two-story 22-ton cylindrical stone mass at the center of Earthwood serves double duty as an efficient masonry stove and the central support column, and it, too, needs to be well-founded. A

Fig. 3.19: At Earthwood, the buttress cage is made of #6 (¾") rebar.

Fig. 3.20:
Twelve pieces of
#6 rebar, as
indicated, were
used to form the
two buttress
cages at
Earthwood.

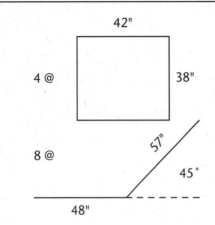

six-foot diameter footing at the center of the building would distribute the load at around 1,600 pounds per square foot, a similar kind of number to the nearly 1,800 psf I'd figured for the saturated load on the perimeter walls. Any settling of the various parts of the footing would be of a similar kind.

Once again, we chose ten-inch-deep forms, but this time we formed with ¼-inch plywood.

Fig. 3.21: Formed with ¹⁄₄" plywood, the footing for the masonry stove is ready to pour.

With an inch of Dow Blueboard™ at the bottom, the concrete depth would be 9 inches. We installed plenty of ⅝-inch (#5) rebar throughout the stone mass footing, criss-crossed and supported as shown in Figure 3.21.

The Post Footings

As explained in Chapter 1, the post-and-beam frame midway between the perimeter of the home and the central mass has seven out of the eight posts of a true octagon. The elimination of one post, made possible by a heavy oak girder, means that in terms of load, the two posts supporting the heavy girder are really carrying a 50 percent greater load than if three posts had been used. Therefore, while the other five posts have footings of three-foot diameter, the northeast and northwest post footings have diameters of three-feet eight-inches.

These post footings were actually done without forms. The lower two-thirds of their nine-inch thickness was simply sculpted out of the compacted sand pad. A disk of Dow Blueboard™ was placed at the bottom of each six-inch deep hole, and the sand sides of the hole were chamfered and wet-compacted just before the pour. Unlike other footings, the concrete was left rough, so that the floor pour would bond well to the cold concrete later.

Pouring the lower part of the post footings was done for safety and convenience. We didn't want seven holes in the pad for workers to trip into during the floor pour. Incidentally, we did criss-cross pieces of rebar through the lower part of the post footings, similarly to the central mass footing.

Concrete Calculations

You've got to know how much concrete to order from the batch plant. I find it most convenient to measure and calculate in feet, or decimal fractions of feet, then multiply according to volume formulas to get cubic feet, and then, finally, divide by 27 to get cubic yards, the unit for ordering concrete. (There are 27 cubic feet in a cubic yard.) For footings, you can take the cross-sectional area of the footing (in square feet) and multiply by the mean perimeter of the footing track (in feet) to get so many cubic feet. For example, a footing 24 inches (two feet) wide by 9 inches (.75 feet) deep gives a cross-section of 1.5 square feet (2 feet × .75 feet = 1.5 sf). Say the mean perimeter (the total distance around the center of the footing) is 140 feet. Multiplying: 140 feet × 1.5 sf = 210 cubic feet. Finally, dividing by 27 gives 7.77 cubic yards. The safe thing in this case is to order 8 cubic yards, and be sure you have some place to use up the extra .23 cubic yards. You might want a pad for a picnic table, for example, or a sidewalk or paving slabs or some other project.

Accurate depth measurements are critical with concrete calculations. In the example above, if the true footing depth is 9.5 inches (.792 feet) instead of 9 inches (.75 feet), it throws the numbers off quite a bit. (2 × .792 = 1.58 sf × 140 feet = 221.67 cf ÷ 27 = 8.2 cubic yards.) That extra half-inch

means that an 8 cubic yard order will fall .2 cubic yards short, which is quite a bit. For this reason, I always average 20 or more measurements around the track to get the true depth, and that averaged number is the one I plug into the calculations. For example, 9.21 inches is .77 feet because 9.21 inches ÷ 12 inches (1 foot) = .77. To get a depth measurement, simply place a straight edge across the top of the footing track, and measure down to the bottom of the track. If you have an inch of extruded polystyrene in the track, measure to the insulation under load (stand on it) or measure before installing it and subtract an inch from your final depth average.

The formula described above will work for any shape of footing perimeter (round, rectilinear, octagonal, etc.) having a regular cross-section. For a cylinder, such as the central mass footing, use the formula for a cylinder: $V(cyl.) = \pi r^2 h$, where π = pi, or 3.14, r = the radius of the cylinder (three feet in our case) and h = the height of the cylinder (nine inches or .75 feet) So: $V = 3.14 \times 3 \text{ feet} \times 3 \text{ feet} \times .75 \text{ feet} = 21.21$ cubic feet. Dividing by 27 gives .785 cubic yards. At Earthwood, this number must be added to our perimeter footings total, as well as the seven little post footings, to get the grand total for the concrete order.

Ordering Ready-Mixed Concrete

For your information: Today's concrete trucks can carry about 10 cubic yards. If you order more than 10 yards, therefore, it will come in two loads. If you order less than four yards, you might be charged a "small load" fee, especially if you are some distance from the batch plant. Ask. And don't leave yourself short by playing too close to the numbers, but have a place to use any overage, otherwise it just makes a mess somewhere.

I have to brag for a moment: In May of 2004, I had a fairly complicated concrete calculation involving a thickened-edge monolithic slab for our octagonal guesthouse, seen in Figure 3.7. An intern was a bit perplexed at the hour or so I spent measuring and computing the volume of the pour, which, coincidentally, came to almost exactly four yards, and that is what I ordered. There was a quantity of concrete left over the size of a small loaf of bread, and I have kept it as a trophy. Even the driver smiled and shook his head, impressed.

But I also had a number of washed clean stones at the ready, and knew I had four 50-pound bags of dry concrete mix in the garage.

For footings, order 3,500 psi concrete, also known as a "six-bag" mix, because six cubic foot bags of Portland cement are mixed with the sand and crushed stone aggregate which make up each cubic yard. For footings, you don't need fiberglass or polypropylene reinforcing, although that is a good option when pouring floors, as we will see in the next chapter.

Be ready for the truck! Have your crew assembled and have all preparations completed an hour before the scheduled delivery time. Last minute prep can take a lot longer than you ever imagined, so have it done way ahead of time, and enjoy a break while you wait for the truck.

The Footings Pour: A Final Checklist

1. Are all the forms braced well enough to resist "concrete blowout?" Besides frequent staking — every five feet on two-by material, twice that amount for plywood – you can pack with wet sand and use tie cleats (see Figure 3.12).

2. Is all the insulation in place? Are all wooden forms exposed to concrete oiled?

3. Is the rebar situated in the lower half of the pour, and properly supported?

4. Are any necessary key pieces made, oiled, and at the ready?

5. Can the truck's chute deliver concrete to all parts of the pour? Wheel-barrowing concrete is … well, boring would be a conservative description.

6. Do you have enough tools for your crew? (A couple of steel rakes, a couple of shovels, straight edges for screeding, flat trowels.)

7. Do you have a crew? It's time to blow the whistle on some strong backs that owe you a favor. For footings, a crew of four sturdy individuals should suffice to draw the concrete around the forms and tamp it in place (see Figure 3.22, opposite).

The Pour Itself

Ask the driver for a stiff mix. Tell him you want a *three-inch slump*. You'll impress him with the terminology, but that's not the point. There are standard metal cones of a certain size, with a handle on the side of the cone, which is placed on a flat surface, wide side down, and filled with fresh concrete. The cone is removed and the "slump" of the cone (the amount the cone collapses) is measured. In the real world, drivers don't carry slump cones around with them, except for expensive engineered projects, but they know that a three-inch slump is a stiff mix.

Why the stiff mix? The driver might even offer to "soup it up" so that the concrete flows around the forms like water.

Nothing doing! You've specified 3,500-pound concrete for the footings, but you'll only get the full strength with stiff concrete, not porridge. The more water (the greater the slump), the weaker the concrete. Maybe it's Friday afternoon. The driver wants to dump this load, clean his truck and hit the pub. But you won't have the strength you've paid for. *Three- inch slump!*

The work itself is easy, but hard, if you get my meaning. Raking, hoeing, and shoveling concrete, especially stiff concrete, around the footing tracks is strenuous, but it's not a highly skilled job. Try to vibrate the concrete down into your footing tracks with your tools, in order to prevent voids. Instruct everyone to be careful about not disturbing the rebar. After one of the sides is poured (or 30 feet of a circular footing), set one or two of your crew to screeding the concrete with a 2-by-4, as shown in Figure 3.23, below. Simply draw the 2-by-4 back and forth quite stiffly by hand, constantly pushing excess concrete in front of you. If too much concrete accumulates in front

of the screed board, shovel or rake the excess away to a place that needs it.

After screeding the concrete, you'll want to trowel it smooth with a rectangular plasterer's trowel. If you have someone on the crew with

Fig. 3.22: Pouring the footings at Earthwood.

Fig. 3.23: The concrete is screeded with a short, straight-edged board.

experience at this, let that person "have at it." It's not particularly difficult, and you can learn how to use the trowel fairly quickly yourself, but experience is nice. The main thing is to keep the leading edge of the trowel up so that it doesn't dig into the concrete … but you'll soon learn about that.

If there are key-way boards to install (as shown earlier in Figure 3.2), they are put in place before the screeding. Screed right over them, but keep them in sight, so that their top surface is the same as the screeded top of the footing. You can trowel to them, too. The key-way boards will stay in place for about a day, then removed. Oil them beforehand.

Removing the Forms

Concrete's rate of setting depends on its strength, ambient air temperature, and humidity. In general, forms are removed the day after the pour, and they are easy to remove if there is a Styrofoam® buffer up against the forming boards, or if you have oiled the boards at any place insulation is not used.

Be sure to clean all of your tools at the end of the job. Usually, the truck driver will offer to help with his high-pressure spray hose that he uses for cleaning his chutes. If not, ask him.

At Log End Cave, it took six of us (one with pouring experience) about two hours to draw and level the concrete – not too bad. The Earthwood footings took around the same time, even though we had the central mass and the bottom half of the seven pillar footings to do, but we were more experienced. In each case, Jaki and I removed the forms the next day.

On Homemade Concrete

The reader may wonder why I do not advise mixing one's own concrete on site, from sand, Portland cement and crushed stone. In response, I want to say, firstly, that, with a name like Rob Roy, I'll squeeze a shilling 'til the Queen screams, but I stop at mixing my own concrete, with the following exception. With small projects of two cubic yards or less, I have actually mixed Sakrete™ (dry pre-mixed concrete) in a wheelbarrow, two 50-pound bags at a time. At our eight-foot square outdoor bathhouse, Jaki and son Rohan mixed 20 wheelbarrow loads – 40 bags – while I poured, screeded, and troweled. I don't mix my own concrete with larger projects because the savings are minimal, especially if you place any value at all on your time. Also, a footing poured over several days (the likely scenario when mixing your own) won't be as strong because of all the "cold joints" between sections.

Chapter 4

THE FLOOR

Not everyone likes or wants a concrete floor. I will not argue. Concrete is hard on the feet and tough on the spine, particularly as you get older. Nevertheless, it is the floor that I am going to accent in this book, for several reasons. First, the concrete floor has heat-retaining mass, a particularly attractive feature when combined with in-floor radiant heating. Second, the concrete floor is fairly inexpensive and easy to do. And finally, you can put all sorts of surfaces on top of the concrete floor, some of which can "soften" it somewhat. Options include: slate, tile, stain, bamboo, cork, carpet, throw rugs and mats, even a wooden floor, such as parquet or hardwood planks.

My friend Mike Oehler takes a different view, touting the earth floor as much kinder to the spinal column. He's right. He likes a low-cost floor consisting of tamped earth, covered with 6-mil polyethylene and carpet, particularly recycled carpet. Again, I cannot take issue with my esteemed colleague. And, if the reader wants to learn about these systems, buy or borrow Mike's *$50 and Up Underground House Book*, listed in the Bibliography.

This leaves us with one more option: the wooden floor supported on floor joists. We'll discuss that, then the concrete slab.

A Wooden Floor

One of our students, Chris Bushey, didn't want a concrete floor in his single story Earthwood house, so here's what he did:

Chris poured his perimeter footings and central mass footings exactly as described in the previous chapter, but, instead of individual post footings, he poured a circular "grade beam" halfway between the perimeter and central mass footings, as per Figure 4.1.

The grade beam is kind of like an internal footing, and serves double duty by supporting the octagonal post-and-beam frame as well as the 2-by-10 floor joists. Like the perimeter footing, the grade beam is 24 inches wide.

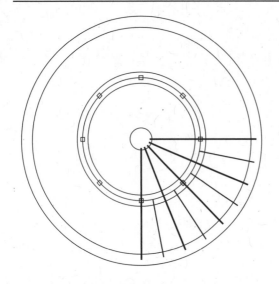

Fig. 4.1: Chris Bushey's footings and floor joist plan.

Over his floor joists, the floor is simply rough-cut "five quarter" (1¼-inch) white pine planking. The intent was always to use this as a sub-floor for a finished floor above, but, after seven years of living in the home, the finished floor is still in the future. The planking has shrunk a little, but the floor is not cold. Chris still speaks of installing a new floor, which would cover all the shrinkage gaps. We discussed all this while I was writing this section, and agreed that if the floor were made "tight" by the addition of finished flooring, then a few floor registers at strategic locations would be a good idea. A fan could also be used to promote air movement. But follow code. Listen:

Chris did cover the sand pad with 6-mil plastic, but there are gaps where it wasn't tightly installed and some moisture seems to move around the plastic. Chris has observed a little moisture damage, but "nothing serious." We discussed what he might do if he were doing it all again. Chris told me that he'd be more careful about installing his 6-mil vapor barrier. Where there is no continuous slab, some building codes require that the plastic be installed up the foundation walls, with all joints carefully taped and sealed. Check with your local code enforcement officer at the design stage.

Chris also thought that an inch of Styrofoam® over the plastic would be a worth-while investment, although he is pleasantly surprised at how warm his floor – just a sub-floor, really – actually is. Finally, he would take extra care to make sure that there was at least three inches of clearance between the underside of the floor joists and whatever is below them, be it sand, plastic, or rigid insulation.

The Bushey home is warm, charming, and comfortable. The wood floor, while a little rough, has worked out well.

THE CONCRETE FLOOR

The considerations for pouring a concrete floor are quite a bit different from those for pouring the footings. This is the major reason that, on a home at least, I prefer doing the two jobs separately, as opposed to the single "monolithic" pour. With small buildings, like our new octagon, I pour the footings and floor at the same time, the "monolithic floating slab," as seen in Figure 3.7.

Preparation

Preparation always takes a lot longer than the pour itself. At the Cave, we poured the footings on July 11th. We thought we'd be ready to pour the floor on July 16th, but were not fully prepared until a week later. There is so much that has to be done. One of the jobs – which I would not do again – was to lay up the first course of blocks. On this home, we used the floor to "foot" the base of the block wall from being pushed in by the earth backfill. If I were building a similar home today, I would use the key-way technique for locking the block wall to the footings, as per Figure 3.2. The blocks, as we were to learn, were also in the way of the power screed's planks. Even with notching the planks to ride the top of the blocks, the rough texture of the blocks made the screed move sluggishly.

The surface-bonded block wall, however, was a great success. We used the technique at both Log End Cave and Earthwood, and it is the main thrust of the next chapter.

Under-floor Drainage

Before doing the first course of blocks, we spread four inches of compacted sand on the undisturbed earth of the excavation for good under-floor drainage. Three days after the footings pour at Log End Cave, we built a sand ramp over the frost-wall portion of the footings with a bulldozer, and pushed and spread two dump-truck loads of sand over the floor area. Using the footing as a visual guide, the operator

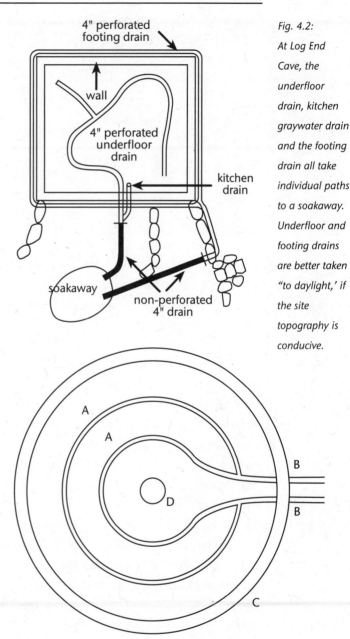

Fig. 4.2:
At Log End
Cave, the
underfloor
drain, kitchen
graywater drain
and the footing
drain all take
individual paths
to a soakaway.
Underfloor and
footing drains
are better taken
"to daylight,' if
the site
topography is
conducive.

Fig. 4.3: A = perforated drain; B= non-perforated drain; C = perimeter footing; D = central mass footing. For a round house, 4" perforated drain tile should be set into the sand so that all parts of the pad are within a few feet of a drain. Slope the drain slightly downwards and take it out above grade. A soakaway is an option on flat well-drained land.

Fig. 4.4:
4" perforated
drain tubing is
set into the
sand pad.

did a good job of spreading and "backblading" the sand to a consistent four-inch depth. The next day, we hired a power compactor to tamp the sand. Jaki watered the sand with a watering can – we had no running water on site – while I ran the machine. Hitting each square foot three times with the compactor results in a solid base for the concrete floor.

With an ordinary garden hoe, we carved tracks in the compacted sand and installed 4-inch perforated drain tubing (sometimes called drain "tile"), as seen in Figure 4.2 for Log End Cave and Figure 4.3 for Earthwood.

The top of the drainage tubing is about level with the top of the compacted sand, as seen in Figure 4.4 taken at Earthwood. You can join sections of tubing with unions or T-connectors made for the purpose. The tubing is a black, ribbed, flexible hose with a white nylon or fiberglass filtration sock to keep fines out of the

pipe. It's a cheap insurance against water problems from below, whatever their source.

At both Earthwood and Log End Cave, we took the drain tubing out to a soakaway in front of the home. The soakaway (sometimes called a "dry well") is a large hole filled with grapefruit-sized stones gathered from an old stone wall. This was a choice predicated upon terrain. A better place to take footing ("French") drains or under-floor drains would be to a point out above grade. Although we have not had a problem, soakaways, in extreme wet conditions, can fill up, and then there is no place for the water to go. Obviously, any drain must run downhill, whether it goes out above grade or to a soakaway. When you take such a drainpipe out above grade, cover its end with a rodent-proof (¼-inch mesh) screen or hardware cloth; otherwise, your drain system might become an attractive chipmunk condominium.

Finally, do not create a U-trap underneath the footings, which can fill with water. The flow must be a positive continuous drain – ever so slightly – down.

EARTHWOOD: INSULATION

In northern climates, put at least an inch of extruded polystyrene under the floor, two inches if you are installing in-slab radiant floor heating. In southern climes, install 6-mil polyethylene (either clear or black is acceptable) under any concrete pour that is not insulated. If fresh wet concrete is poured on sand without plastic or rigid foam, the sand will suck the moisture out of the concrete so fast that it will "get away" from you (set much too fast), leading to a panic to get the floor troweled before the concrete sets. I know this because I committed the error myself at our summer cottage's monolithic slab, and, event-ually, I had to hire an industrial grinder to smooth the floor. I'm here for you, and happy to share my mistakes.

Figure 4.5, taken at Earthwood, shows the insulation in place, and other details. If you are doing in-floor waste plumbing, under-floor electrical conduits, or combustion air inlets, all discussed below, the insulation goes in after those systems.

IN-FLOOR RADIANT HEATING

Several times, I have mentioned in-slab heating. We don't have it at our home, but, without exception, everyone I know who has in-floor radiant heating is crazy about it. The system consists of special plastic (cross-linked poly-ethylene, usually) tubing, designed for its resistance to concrete, woven throughout the area to be heated, usually on 12-inch centers between parallel lines of run. Normally, the home is divided into two, three, or more heating zones, which can be separately controlled by thermostats. The heat comes from water that circulates through the concrete by way of electric pumps. Fuel can be gas, fuel oil, electric, or even wood.

My neighbor, Steve Bedard, installed his own tubing prior to having his floor professionally poured. Following installer's recommendations,

Steve placed two inches of extruded polystyrene on the sand prior to pouring the concrete. On top of this, he laid out 5-foot-by-10-foot sheets of "six-by-six-by-ten-by-ten" wire mesh. This cryptic designation means that the mesh is 6-inch-by-6-inch grid, with #10 wire in each direction.

Working alone, Steve ran his tubing back and forth along the grid at one-foot (12-inch) centers, making a loop at each end. The tubing is fastened about every four feet with a tie-wire, easy to install with an inexpensive wire-tying tool made for the purpose. See Figure 4.6, below.

Other special right-angle sleeves are available to bring the tubing up safely (without kinking) through the concrete and on to the area where the pump and furnace are located. See Figure 4.7.

The actual mechanics are best left to a professional installer, but if you are of a practical bent, like Steve, and want to save some money on installation, see if you can work a deal with the professional installer to do the tubing yourself.

Fig. 4.5: Extruded polystyrene insulation and wire mesh are installed. The two buttresses are formed. We're ready to pour!

Mesh or Fiber?

The "six-by-six-by-ten-by-ten" wire mesh can be seen in Figure 4.5, where we used it as concrete reinforcing. The sole purpose of the mesh in our case was to hold the concrete together when it cracks. Notice my use of the word "when," not "if." Concrete cracks, period. The mesh is there to prevent an elevation change between one section and another when it cracks. (Yes, I've seen one or two small pours without cracks, but most pours of any decent size have hairline cracks in them, which is not a problem as long as the slab is reinforced.)

Personally, I no longer use the cumbersome wire mesh, which requires a designated person, during the pour, to constantly lift it into the concrete with the tines of a metal rake. Instead, I now order my concrete reinforced with millions of little polypropylene fibers. The fibers add a bit to the cost of the concrete, but eliminate the cost of the mesh and the aggravation that goes along with it. The fibers do the same job as the mesh: They hold the concrete together when it cracks.

If reinforced concrete is not offered at your local ready-mix plant, use the mesh. A tip if you do: use the 5-by-10-foot sheets, not the rolls of wire, which have a "memory" and are very difficult to lay down flat. In either case, lap the edges of two adjacent pieces by a full square of the grid, six inches.

Fig. 4.6: Steve Bedard fastens his polyethylene tubing to the wire mesh with a special rotating hook made for twisting tie bands.

You could have them check your work before the concrete is poured.

UNDER-FLOOR PLUMBING

This book is about earth-sheltered housing. I am neither a plumber nor an electrician, although I have done my own work on parts of these systems at the four houses I've built, particularly waste plumbing. My commentaries on these systems are based on my own experience in earth-sheltered housing and are only meant to supplement plumbing and electrical manuals, not substitute for them. I am pleased with the way our waste plumbing has performed, but not everything I've done has been "up to code," as related below.

Some wise advocates of owner-building have cautioned against installing waste plumbing under a concrete floor, and with good reason: If something goes wrong under there, the first tool you're going to need is a jackhammer. This warning might just be enough to push you towards consolidating your plumbing in as small an area as possible – bathroom back-to-back with the kitchen, for example – and putting all of the plumbing under a wooden floor, something like Chris Bushey's, already described. Waste plumbing is often referred to as the DWV (drain/waste/vent) system.

Nevertheless, I continue to install DWV plumbing under the slab, although I do it very carefully. In fact, I prefer to do the job myself, slowly and quietly, than to sub it out to a plumber, and cost saving is only my second reason for making this choice. The first reason is that *this is my home*. I have a very real stake in making sure that the system will not fail.

At Log End Cave, I used schedule-40 plastic PVC (polyvinyl chloride) pipe below grade. At Earthwood, I switched to cast iron, but I now consider that to have been a mistake, and would not do it again. The reasons: Cast iron is expensive, brittle, difficult to work with, and requires the hiring of special cutting tools. And, my view now (supported by others "in the know") is that the modern schedule-40 plastic pipe (either

PVC or ABS [acrylonite-butadiene-styrene], as we used at our summer cottage) will outlast the cast iron, which is prone to rusting and other damage. For these reasons, cast iron is seldom used anymore, and I'm not even going to discuss it here, except to warn you that a house drain running to a municipal sewage system may require the use of cast iron by local code.

George Nash, in his excellent *Do-It-Yourself Housebuilding*, to which I often refer for technical information for conventional construction, points out that ABS "is considerably stronger and has both greater impact and thermal resistance than PVC pipe." It used to be that you couldn't mix PVC with ABS, but the advent of "universal glues" has now made this possible, if it becomes necessary.

The Cave DWV system (see Figure 4.8, below), was extremely simple, and the bathroom fixtures consisted of a toilet, a bathtub/shower, and a sink.

Fig. 4.7:
Four different heating zones come together at the edge of the building. At tight 90-degree bends, pieces of rigid plastic fit over the polyethylene tubing to prevent kinking.

Fig. 4.8:

The simple under-floor plumbing at Log End Cave. Key:

1. Toilet receptacle unit

2. 1½" to 3" right angle join, receiving tub, vent and basin,

3. 4" sewer pipe, or "house drain"

4. 3" to 4" expander

5. 3" PVC pipe

6. 3" PVC right angle join

7. 3" PVC 45-degree angle

8. cleanout with 3" threaded cap

9. wooden cover

10. vent stack to exterior

11. tub drain

12. 1½" PVC right angle

13. 1½" PVC pipe to basin.

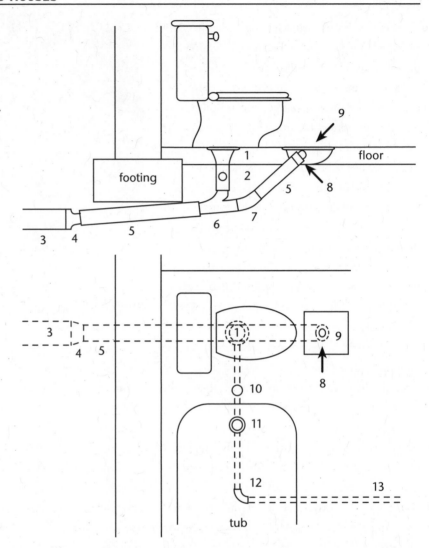

All fixtures entered into a 3-inch PVC schedule-40 "house drain," which expanded outside of the house to a four-inch line all the way to the septic tank.

Our toilet ran straight into the 3-inch PVC house drain. The other fixtures used 1½-inch PVC outlet pipes, which joined to the 3-inch house drain as seen in the diagram. Washing machines require a 2-inch outlet line, as do vent stacks. In large houses, with lots of plumbing, the code-required vent stack might need to be 3-inch pipe or even larger. The vent acts as a pressure equalizer.

A toilet flush needs to flow rapidly down a pipe and out through the house drain to the septic tank. But if air can't rush in behind the flush – air from a vent – then the flush is trying to create a vacuum behind it. Nature, it is said, abhors a vacuum – at the very least, it makes vacuums difficult to exist – and so the toilet will run sluggishly without air from a vent to easily fill the vacuum. Also, a vent stack provides a rooftop escape route for septic tank odors. What we did worked, but code and standard practice both dictate that the vent stack be placed on the septic-tank side of the toilet.

The DWV system was only slightly more complex at Earthwood, because the home had two bathrooms, but most of it can be seen in Figures 4.9 and 4.10. The main differences were that we used cast iron plumbing (not recommended) and something called an "atmospheric vent," also known colloquially as a "pop vent."

Just as a check valve only allows water to flow one way through it, a pop vent only allows air to move in one direction. I think of it as a check valve for air. At the advice of a plumber friend, I used it as a substitute for the vent stack at Earthwood, because I knew that any penetration through the waterproofing of an earth roof – such as a vent stack – is an open invitation for leaks. The pop vent is used at the same location as a vent stack, and does the same job. By letting air in, waste can flow easily, as no vacuum is being created. By not letting air out, no odor back-drafts into the home. Our pop vent is actually

Fig. 4.9: Underfloor plumbing at Earthwood. The top of the cylindrical "red cap," mentioned in the text, is set at finished floor level, which is the same as the top of the footings.

hidden in a wall in the downstairs bathroom, and has worked perfectly for 25 years. We have never had blockage or odor. The only time the system has worked sluggishly is in the spring when the drain field was saturated, but an ordinary code-approved vent wouldn't have helped in those circumstances in any event.

The pop vent worked so well that I used one again at our summer camp, and for a similar reason: I did not want to violate the wooden shingle roof of our geodesic dome upstairs. That vent, too, has never failed. This begs the question: "Why doesn't everyone use atmospheric vents instead of the cumbersome expensive vent stack through the roof?" The answer is, simply, that plumbing codes don't allow it. Why? I don't know. But I mention atmospheric vents because they may provide a solution for some owner-builder with similar concerns to ours, and similarly untroubled by strict plumbing codes. Be informed, but break rules at your own risk.

Fig. 4.10:
Dennis Lee
checks the slope
of the
underfloor
drains.

Installation Tips for Underfloor Plumbing

Plan the plumbing so that all traps are accessible. In the case of a bathtub or shower on the first floor, this will usually necessitate "boxing out" a small area so that concrete cannot get in around the trap or drain. An 18-inch square box made of 2-by-6 material can be seen in Figure 4.9. It provides access to our shower drain and trap. Another good reason for the box is that – thanks to the ubiquitous Mr. Murphy of the famous law – it is very difficult to get the drain in exactly the right place with respect to adjacent walls, which are still nonexistent at this point in time. Within the box, you can still make final adjustments to a drain location at the time of shower- or tub-base installation.

Your DWV plumbing must be set firmly in the compacted material of the pad, so that no settling can occur. (This is another reason that I prefer coarse sand to crushed stone as pad material: it is easier and safer to set the pipes into.) The correct slope for "horizontal" DWV pipes is between ⅛-inch and ¼-inch per foot, towards the septic tank. I like 3⁄16-inch, a middle ground. Using a 2-foot level then, as seen in Figure 4.10, the slope will drop ⅜-inch over its length; with a 4-foot level, the drop is ¾-inch.

Know the exact location of your plumbing appliances with respect to walls. With a rectilinear shape, like the Cave, you have the corners as reference points. With a round building, you need those eight external reference stakes, discussed in Chapter 3. At Earthwood, we poured the lower half of the seven pillar footings at the time of the footings pour. We found it convenient to place an 8-inch-square piece of Styrofoam® at the exact internal post locations, based upon the building's center and our external location stakes. With these "mock" posts precisely in place, it was an easy matter to draw all pertinent walls in the damp sand. Armed with a scaled floor plan, the location of waste pipes can be determined by measuring.

Measure and cut the ABS pipes, being sure to allow for the space taken up by elbows, Y's, T's, and other connectors. Fit your system, unglued, in the prepared tracks. When everything fits right, mark the pipes and fittings with a pencil line (so you can duplicate the configuration), then glue the unit together, one connection at a time. Get a good book on plumbing or seek advice from someone experienced. Gluing plastic pipe is easy, but you only get a couple of seconds to get those pencil lines to line up before the joint is frozen. There are a lot of good tips on the side of the glue can, too, such as deburring joints and cleaning them with a compatible "etching" solvent before joining them.

Pack sand around the pipe after it is set in place and the grade is checked. I use the side of a sledgehammer for this purpose, being careful to tamp the wet sand and not the pipe itself.

Before capping off all the openings, test the system! Pour water in all the openings and make sure it comes out the end of the house drain, briskly. (Too much slope is not a good thing, though; the liquids could outrun the solids.)

Leave all waste pipes sticking up at least a foot and wire them to an adjacent stake. Carefully cap the tops of all pipes so that sand, concrete, or children's small toys don't find their way into your DWV system.

A "toilet receptacle" is a standard plumbing fitting to which the toilet is later bolted. The top of it is set exactly at floor level, which can be determined with a long straight-edge – your screed board, perhaps – running from one footing to another. Or use a laser or contractor's level. The receptacle must also be at the correct distance from the wall the toilet rests against. Use the "roughing in" distance for the actual toilet you are planning to use (usually but not always 12 inches),

measured perpendicularly from the wall with a square. The toilet receptacle, unlike the shower drain, is supported by concrete, so it has to be in the right place the first time.

Cover the toilet receptacle tightly to prevent concrete going down into the DWV pipe. Cast iron receptacles come with a "red cap" made for the purpose (seen in Figure 4.9), but find out what is available at the plumbing supply house to cover the receptacle you have purchased. At the lake cottage, a square of sticky Bituthene® water-proofing membrane served the purpose nicely.

Again, to thwart Murphy's Law, place a cleanout at a strategic position to clear the house drain of any blockage. I have installed them at every home I've built and have never had to use any of them. But if you don't put one in, you will eventually wish you had. Cleanouts for 3-inch and 4-inch pipe, with a threaded cap, are common items. Provide access to them by having them stick out of the concrete on a 45-degree elbow in the direction of flow. Access can be hidden by boxing in the cleanout, as we do with the shower and tub drains. At Earthwood, we have installed a varnished wooden access cover to get at the cleanout, which is entirely below floor level. See Figure 4.11 and refer back to Figures 4.8 and 4.9.

Under-floor Electrical

A plumber friend built a one-story version of Earthwood at about the same time we built our home. In addition to under-floor plumbing, he also ran under-floor conduit for his major

electrical lines, popping up at strategic locations where internal walls would be located. This is a good way to bring electricity from the service entrance to circuits that are quite a distance away. Obviously, a lot of careful planning and measuring has to be done to get the conduits to exactly their right places within a framed wall. But the process is not really so different from the under-floor plumbing, already described. Ask at the electric supply dealer for a code-approved conduit for use under concrete.

Even if you are hiring an electrician to wire your house, you can save money by installing the under-floor conduit yourself. Go over the circuit plans with your electrician, to make sure that you are running the conduit to useful places.

Plumbing and electric are two trades that a lot of owner-builders feel better about subcontracting out. But savings are possible if you can save the tradesman some traveling time and donkey work.

UNDER-STOVE VENTS

Many owner-builders like to heat with wood, and woodstoves need a source of combustion air whether they are massive masonry stoves like the one at Earthwood, or traditional freestanding models as we used at Log End Cave. The best way to provide air is by a direct vent to the exterior. Without this positive vent, the stove will suck its required air from around doors and windows or other leaky places. Or the stove will not function properly; it will "choke." At Log End Cave, we ran 4-inch non-perforated flexible tubing from the exterior, under the footings and floor, and up under the two woodstoves. The inlet is covered with ¼-inch mesh hardware cloth to discourage varmints. Some advocates of this system say that the cross-sectional area of the incoming air pipe should be equal to the diameter of the stovepipe. I can see some logic in that, but our system at Log End worked well. Maybe it would have worked even better with a 6-inch air tube.

When some people hear of this vent, their first thought is that they don't want to bring this cold air into the home. But this is not the case. If you bring the air to a location close to the stove's firebox, the stove will get air (or create a draft) from its easiest source, which is the incoming vent pipe. Now, for another of "Rob's mistakes that the reader can easily avoid." I took our 4-inch non-perforated air vent under the footings, creating a U-trap, not unlike the one under a kitchen sink. After a while, condensation filled the U, and we had to pass a towel back and forth on a rope until we'd bailed out the trap. Design your system so that any condensate runs out of the air vent to the outside.

The masonry stove at Earthwood gets its air from a number of "earth tube" inlets around the perimeter of the home (see Chapter 11), but an excellent alternative would be to bring a 6-inch-diameter combustion air supply vent under the slab and have it come up next to the firebox location. Later, when the masonry stove is actually constructed, that air vent can make a right-angle turn and supply direct air right into the side of the firebox. A damper can be installed in the line so that the incoming air to the stove can be stopped when it's time to close the stove down.

With a tight home such as an earth shelter, it is particularly important to provide an easy, non-drafty source of combustion air, and you need to install it before the floor is poured.

Log End Cave: The Fence

The first step during the actual pour will be to flatten the concrete at the correct elevation, a process called *screeding*, described below. Oftentimes, screeding is done by drawing a long straight plank back and forth across the concrete, supported at both ends by a footing or grade beam. But, with most homes, the screeding span will be too great without including one or more intermediate "fences" to support one or both ends of the screed board. At Log End Cave, we ran a fence down the middle of the narrow dimension

of the pour, to cut the screed span in half, as seen in Figure 4.12. The fence "posts" are 2-by-4 stakes driven firmly into the sand pad. The top fence "rail" is simply a straight 2-by-4 screwed to the stakes, such that the top of the rail is at the same level as the top of the footings. (Or, in the unusual case of the Cave, level with the top of the first course of blocks.) Make sure that the tops of the stakes are all below the top edge of the fence, so that they don't cause troublesome "bumps" in the screeding process.

After the pour is completed, the fence is removed and voids are filled with concrete. All evidence of it is obliterated by troweling.

Fig. 4.12: Pouring the floor at Log End Cave. Note 2 × 4" "fence" on left, screeding and (at top) bullfloating.

*Fig. 4.13:
Check floor
depths at 24
locations prior
to ordering
concrete.
Average the
depths. A small
error on floor
thickness
impacts
calculations
tremendously.*

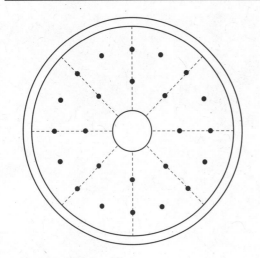

Concrete Calculations ...

... are no more difficult with the floor than with the footings. The same formulas for volume are at play. With Earthwood, we think of the floor as a disk, which is nothing but a very thin cylinder (Volume =$\pi r^2 h$). We calculate a volume for the entire disk, within the external perimeter, and then subtract the portion of the floor already taken up by the round central mass footings.

With slab calculations, even more than with the footings, the actual thickness of the slab (h in the formula) has a tremendous impact on the volume result. The best way to get an accurate depth figure is to take 24 measurements at the locations indicated in Figure 4.13, below. Place your screed board (or other straightedge) from the center mass to the perimeter at the various locations and, with a rigid ruler or yardstick, take the 24 readings to within a sixteenth of an inch. Change standard measurements to decimal, so 3½

inches = 3.5 inches, 3⅝ inches = 3.625 inches, and 4³⁄₁₆ inches = 4.187 inches. Maybe your final average is a number like 3.77 inches. That is your height (h). A difference of a half-inch here can mean a couple of cubic yards of concrete, one way or the other.

THE FLOOR POUR: A FINAL CHECKLIST

1. Can the truck's chutes access all parts of the pour easily?
2. Is the six-by-six-by-ten-by-ten wire mesh installed and do you have a designated "mesh puller-upper" armed with a metal rake, ready to keep lifting it into place?
3. Or, if you're not using the mesh, have you ordered fiber-reinforced concrete?
4. Have you got a good crew of at least six people? Use those experienced bodies from the footings pour and try to find somebody who can run a power trowel. If you've never run one yourself, it is probably best to not begin your learning curve on your own floor. For the floor, I recommend that you hire one experienced person to direct operations and, perhaps, run powered equipment.
5. Has all your under-floor piping (plumbing, drain tubing, air supply vents) been tamped into place? Did you test your plumbing with water?
6. Do you have all the necessary tools for your crew? Three steel rakes and three

spade-type shovels won't go astray. Have a long enough screed board on hand, at least a foot longer than the span you need to screed. Fasten a vertical handle to each end of it, such as a 16-inch length of one-by-two. Hire a bull float and a power trowel. A power screed is a nifty tool, too, as seen in Figure 4.15 below, but, like the power trowel, it should be operated by someone with experience.

7. If spans are too great for a reasonable length of screed board, have you set up a strong "fence" to shorten the span, as per Figure 4.12 on page 89.

LOG END CAVE: POURING THE FLOOR

Again, as with the footings, you want a fairly stiff mix to get the full strength you are after. Order 3000-pound (five-bag) mix.

Experienced flatwork contractors will direct the driver where and how much concrete to pour, but, for do-it-yourselfers, most drivers are quite good at placing the right amount of concrete in front of the screed board. But the responsibility for the call is yours, not theirs.

If you are using the mesh, make sure the designated "mesh puller-upper" keeps lifting it into the concrete, off the Styrofoam®, but not to the very top of the pour, where it will get in the way of troweling. I recommend the fiber-reinforced option.

Work one section at a time, between the fence and the footings, for example, as seen in Figure 4.12. As soon as you've got a fair shot of concrete placed, two people should get on it with the screed board. (Worth repeating: strong handles at either end of the 2-by-8 straightedge make the screeding much easier.) The two crew members at the ends of the screed board need to get into a rhythm, like two people working a large crosscut saw. Another helper (with a rake) can draw extra concrete away from the front of the board, to make it move easier, while yet another crew member (with a shovel), can throw a little extra concrete in depressions, as needed. The crew chief (you or your hired gun) tells the driver when and where to place another "shot" of concrete. (See the Sidebar, Steve's Pour, to see some of the tricks that pros use on the job.) I was fortunate to have a contractor friend, Jonathan Cross, on hand to keep everything running as well as it did.

Learn from my experience …

At Log End Cave, the concrete truck arrived about noon. We had rushed all morning, beginning at 5 a.m., to get everything prepared, and were still leveling off the fence while the truck was mixing. This is cutting things too close. Be ready the day before.

The first half of the slab went well, and then we had a bit of a scare. We had over an hour's wait for the truck to get back with the rest of the concrete. We used the time to catch up on

Steve's Pour

In the autumn of 2004, my neighbor Steve Bedard poured concrete walls and a concrete slab at his new home near Earthwood. He graciously phoned me each time they poured and I would record the event with my cameras.

Steve's professional crew used a power screed on the concrete without benefit of supporting fences or adjacent footings to guide the depth. This was accomplished by first pouring several flat pads of concrete, the tops of which were checked for proper grade with a laser level, and marked with an "X." Then concrete was poured between these benchmarks and the power screed leveled this new concrete to the same grade as the benchmarks. After a couple of concrete levies were made in this way, at the proper grade, more concrete was poured in the large in-between sections, and the power screed was used perpendicularly to its first use, getting the

Fig. 4.14: The beam of a rotating laser level, 25 feet away, hits a grade stick. A benchmark on the grade stick shows if the concrete pad is high or low. Adding concrete or taking it away brings the pad to finished floor level.

screeding, floating, and troweling, and for lunch, but, because there was no plastic or Styrofoam® beneath the concrete, the first load was setting rapidly with no sign of the second load

Concerned about stiffening concrete, we watered the edge of the first pour and inserted 16-inch sections of rebar eight inches into its edge to tie the two pours together. The second load arrived just in time to save a nasty looking "cold joint" between the two pours.

Log End Cave's slab was 17 cubic yards. Drawing all that concrete with rakes, screeding it, battling with mesh and a power screed that didn't slide very well over the course of blocks, working around plumbing ... well, it was a long day. By five o'clock, all our help was able to go home,

right amount of concrete distributed uniformly. I was impressed with this crew, who do this sort of work every day, but I'm sure that the skill level of running the power screed is considerable, and I would not advise inexperienced owner-builders to attempt it.

After screeding, the next process is smoothening the concrete with a "bull float," which also serves to bring the fines to the top of the concrete for ease of troweling. The bull float is a large flat masonry blade, made of aluminum or magnesium. Extension handles enable you to reach 15 feet or

Fig. 4.15: One man operates the power screed while another bullfloats right behind him.

more onto the slab so that concrete can be floated fairly flat and smooth. The bull floating process can be seen in Figure 4.15. The bull float itself is commonly available at tool rental stores.

Finally, Steve's professional crew power-troweled his floor to a nice smooth finish. (Figure 4.17, later on, shows a power trowel in use.) Again, this is a skilled task, which some capable owner-builders will be able to learn, while others will get hopelessly mired in their work.

except Jonathan, who had plenty of experience with the power trowel. He stayed on for supper, checking the concrete frequently to see if it was stiff enough to walk on and run the power trowel. When the time was right, he started the machine and guided it, with its four rotating blades, across the floor. I sprinkled water and/or dry Portland cement powder on as needed to produce a

smooth floor. We finished after 10 p.m., by car headlights. It had been a long day, longer than it should have been. But, thanks to Jonathan's experience, the slab came out pretty well.

The three morals to this story are: (1) Be ready for the pour a good hour before the truck arrives. (2) Never pour without at least 6-mil plastic – or extruded polystyrene in northern climes – on top

EARTHWOOD: PREPARATION FOR THE FLOOR

One of the main differences in the floor pour at Earthwood, compared with Log End Cave, was that, prior to the pour, we placed an inch of Dow Styrofoam® Blueboard™ over the sand drainage layer. Not only does this insulation make the tons of concrete much more valuable as mass heat storage, but its closed-cell nature works about as well as plastic sheeting in preventing rapid wicking of water from the concrete to the sand pad, thus delaying the set to a more comfortable working time. Concrete shrinkage and cracking is reduced at the same time.

As we had already placed the insulation under the seven pillar footings, we did not extend additional insulation over these locations. The rough concrete of the bottom half of these pillar footings would become a part of the floor pour at these locations.

Plumbing preparation has already been described above.

EARTHWOOD: FORMING AND POURING THE BUTTRESSES

During the footings pour at Earthwood, you may recall, the heavy reinforcing bar cages for the buttresses were left extending out of the concrete. Pouring the buttresses themselves had to wait for another time, and the floor pour was that perfect time. We formed the buttresses with recycled 2-by-6 tongue-and-groove silo boards, seen in

Fig. 4.16: We formed the buttresses with recycled 2 × 6" tongue-and-groove silo boards.

of the sand to slow the set. And (3) Specify two trucks, maybe 45 minutes apart, so that there is no waiting for the second half of a large pour.

Figure 4.16, and fastened the boards together with 16-penny forming nails, also called duplex or double-headed nails. The interior of the box thus created was about 43 inches high, 45 inches long, and 12 inches wide, although, as already stated in Chapter 3, I would go 16 inches wide if doing it again, to match the wall thickness. All rebar was about three inches from any edge of the concrete buttress.

The form was not fastened down during the pour, but care was taken to hold it in place until a foot or so of concrete had been installed to anchor it, thus preventing a lifting of the wooden form. We actually poured the buttresses first on "floor day," so that people wouldn't be stepping on the freshly poured floor to do the buttresses later.

We used a long metal rod to tamp and vibrate the concrete into the form to eliminate voids. Be careful not to over-vibrate, as this can put a tremendous pressure on the strongest of forms, and concrete "blowout" is no fun at all.

EARTHWOOD: POURING THE FLOOR

In addition to the buttresses and the floor itself, we also poured our little 10-foot-diameter monolithic thickened-edge floating slab for the sauna. All of these volumes must be added together to come up with the total amount of concrete required.

Again, we knew it would take two trucks to provide the necessary concrete, but, this time, we made sure that they would come one right after the other. Again, I used six-by-six-by-ten-by-ten

wire mesh, because, in 1981, fiber-reinforced concrete was still not common in our region of New York. Now I use the reinforced concrete instead of mesh on all slabs.

We were able to screed the concrete with a straight 16-foot 2-by-6 plank, which covered the 15-foot span between the central mass and perimeter footings. A 2-by-8 screed, with handles, would have been even better. Because the inner diameter of Earthwood is only 36 feet, we could pour the floor right up to the footings without the use of an expansion joint at the outer edge. At 40-foot diameter – or any rectilinear pour with a 40-foot dimension – you will need an expansion joint to prevent outward pressure on the footing in case of heat expansion. This expansion joint can be a ½-inch piece of Homosote™ board (a

Fig. 4.17: Jonathan Cross power-trowels the slab at Earthwood.

fibrous insulation board made from pulp) or even a half-inch of polystyrene insulation.

In those days, we were still in the "wire mesh mode," so one of our crew, with a rake, was designated to keep pulling the six-by-six-by-ten-by-ten mesh off of the Styrofoam® (where it is doing no good) and up into the 4-inch-thick concrete. Keep mesh an inch or two off the bottom, but be particularly careful that it does not wind up near the top of the pour, where it might get in the way of power troweling.

In the actual event, we had about three-quarters of a yard of concrete left over after the floor, sauna, and buttresses were poured. My assistant, Dennis Lee, and I quickly formed out a trapezoid-shaped greenhouse slab on the south side of Earthwood, using two-by-four studs as forming boards. It came out okay, but it should have been formed out ahead of time, as the Styrofoam® and wire mesh were applied rather hurriedly. Always have some useful space formed out to receive any leftover concrete.

During hot weather, moisten the slab frequently during the first 24 hours to help retard the set of the concrete. Curing in this way will lessen the size and frequency of cracks.

Chapter 5

External Walls

A list of wall choices for earth-sheltered housing would include surface-bonded concrete blocks, conventionally mortared blocks, stone masonry, cordwood masonry, poured concrete, pressure-treated wood walls, rammed-tire walls ("Earthships"), and Mike Oehler's PSP method. All of these methods have been used successfully and I would be remiss to say otherwise. However, it is my strong view that surface-bonded concrete block walls are the most appropriate for the inexperienced owner-builder, so that is the method described in detail in this book. Here are some comments on the other methods.

Conventionally Mortared Concrete Blocks

Millions of basement walls in the US are made from concrete blocks mortared up with a ⅜-inch mortar joint. Most of them have served the purpose fairly well, but not a few of them have suffered from failure because of the weak bond between mortar and blocks, characterized by horizontal and vertical cracking along the mortar joints. Seldom do the concrete blocks themselves break. Once a crack develops in the wall, water is likely to find its way in. Conventional block wall basements are normally "damp-proofed" (not waterproofed, we will discuss the difference in Chapter 7) with either a cement-based product such as Thoro Corporation's Thoroseal™ or one of the various goopy black tar or plastic coatings. These products work quite well ... until the block wall cracks. In all cases, avoid lightweight aggregate blocks, sometimes known as "cinder blocks." Always use genuine full-strength concrete blocks.

The tensile strength of the inner surface of a conventionally mortared wall is quite a bit less than that of a surface-bonded block wall.

Finally, laying concrete blocks is a skill that comes from practice. It is very much more difficult to do, and do well, than surface-bonding.

Stone Masonry

I like stone masonry. I find the work to be quite therapeutic. But it is extremely time consuming.

Steve Bedard's Pour

If you are not inclined to build your own surface-bonded walls, poured concrete walls are certainly a viable option. Poured walls can be done quickly by professionals and they will be strong. Choose your contractor by price and by reputation. Cheapest is not always wisest. Steve Bedard chose an excellent contractor, specializing in concrete work. Despite the tremendous quantity of concrete used in the walls of his home (over 60 cubic yards), the work went like clockwork, and Steve ended up with a good product. Steve is a very handy fellow, himself, and worked with the crew, just as hard as any of the paid workers. He figured that it was his house, and he wanted to know exactly how it was put together. Steve didn't get in the way, but, rather, stayed useful handling the chute, acting as a "go-fer," and making any "executive decisions" that came up (after consult with the pros, of course).

Fig. 5.1: Strong forms retain the concrete, which is delivered by chute, directly from the truck, or by hopper, controlled by the operator of the articulated boom.

Forming – and its strength – is critical. The pros have a quantity of strong wooden

A lady neighbor built her own stone basement using the slip-form method of construction developed by Flagg and popularized by the Nearings and others. Although slip-form stone construction is faster and easier than freeform, it still took a summer of pretty much constant work to build her small basement. To properly waterproof a stone basement, make the outer surface smooth with one or – if necessary – two coats of cement plaster parge, and then apply one of the waterproofings described in Chapter 7.

Unless you are a skilled stonemason, slip-form is the way to go. See the Bibliography for reference books on this method.

Stone masonry is for someone with lots of stones and lots of time. My neighbor – and the Nearings – had both. Personally, I'd sooner go with the surface-bonded blocks, the main thrust

forming components that they can use to build up any required size and shape. Frequent heavy metal banding and corner angle iron braces are all bolted to each other, eliminating any chance of blowout, even during concrete vibration, a procedure that puts additional stresses on the forms.

Placement and proper wiring of rebar (to keep it in place during the pour) are critical skills that come from experience. Frequent cross-ties stop the forms from spreading, and become a part of the wall when the forms are removed. Engineer's plans will specify quantity and size of metal reinforcing. Pros like those Steve hired know these specifications as well as a framing contractor knows how to make stud walls.

Fig. 5.2: Strong forming is critical to pouring concrete walls. Note frequent cross ties that keep the forms from spreading.

I watched as truck after truck came and distributed their loads amongst the various framed walls on site. Many of the forms were filled directly from the concrete chute attached to the truck. Other walls, hard or impossible to reach with the chutes, were poured from a hopper, carried to the required location by a large articulated boom.

Again, these are professionals. Don't try to do this at home.

of this chapter. They are infinitely faster, much brighter (unless you put another white plaster coat on the interior of the stone wall), and more regular for continuing construction above grade or for applying waterproofing and insulation.

CORDWOOD MASONRY

I only mention this as an option because prospective cordwood builders often ask me if it is

suitable below grade. In fact, Jaki and I originally intended to use cordwood masonry below grade at Earthwood, and began to do so. We used very dry hardwood log-ends instead of our usual choice of white cedar, and encountered the problem of the wood swelling when rainwater collected on our slab, necessitating the tearing down of our beautiful walls … twice. Our intent was to return the outside of the wall to a true smooth cylinder with two coats of cement plaster, and even did this successfully on a small scale at our sauna, but, once we discovered that the dry hardwood was a bad choice, we fell back on the tried and true method of surface-bonded blocks. Now I tell students that, even with log-ends that will not swell, I would be disinclined to use cordwood masonry below grade.

Like stone, it is much slower than blocks, and, again like stone, cordwood masonry is a light-absorbing medium. Finally, in my view, one of the main reasons for cordwood's longevity in its normal use above grade is that breathes so well along its end grain. If one end is sealed off with a waterproofing membrane, this "breathability" is compromised.

POURED CONCRETE

Poured concrete walls are strong and the work goes quite quickly. (See sidebar showing Steve Bedard's wall pour.) However, it is not a method I recommend for inexperienced owner-builders – for three reasons: (1) There is a very real danger of "concrete blowout" during the pour. Forming must be very strong to support the tremendous weight of concrete. My strong advice is to subcontract this job out to professionals. An owner-builder near us poured his own walls, and suffered a concrete blowout – yards of concrete wound up horizontal instead of vertical. This tends to spoil your day. (2) It is difficult to get the reinforcing rod (rebar) correctly placed. (3) Because walls should be poured by professionals,

the actual out-of-pocket cost will be two to three times the cost of doing it yourself with surface-bonded blocks. Another drawback with poured concrete is that the walls take quite a long time to transpire all of their embodied water. With a waterproofing membrane on the exterior, the walls can only dry to the interior. It takes a year or more to bring the humidity in a poured wall earth-sheltered space down to comfortable levels. But it will happen, eventually.

PRESSURE-TREATED WOOD FOUNDATIONS

Some contractors, and even owner-builders, have built earth-sheltered walls of pressure-treated (PT) framing and PT plywood. I don't like them, full stop. They can be made strong, and many have been successfully built. But I have also heard horror stories of failure. They have to be engineered carefully to resist the pressure of backfill. But my main objections are these: (1) Walls are insulated in much the same way as ordinary above-grade stud walls. With most of the other heavy wall systems, it is possible to place the rigid insulation correctly on the exterior of the mass. The mass acts as a giant capacitor for heat storage, helping to maintain steady temperatures in the home. There is almost no mass working for us with PT walls. (2) Pressure-treated materials are full of chemicals, chemicals designed to discourage deterioration in the wood. Whatever is used to discourage fungi and rot is not likely to be very healthy for the human occupants, either, and all of the off-gassing of these chemicals is into the living space. Some of the really nasty chemicals have been banned in PT wood, but I'm not convinced that their replacements have been around long enough to assure safety. It may be possible to stop off-gassing of chemicals with some other "impervious" coating, but I am in sympathy with healthy-house proponents and "natural builders" to the point that I will not use any PT material inside the home. (3) Finally, one of the worst exposures to the dangerous chemicals is during cutting and handling at the time of construction. I am still guilty of using PT material, where appropriate, on a home's exterior. Always wear proper eye and nose protection when cutting PT lumber and gloves when handling it.

RAMMED-EARTH TIRE WALLS

Michael Reynolds has popularized the "Earthship" style of home, and they seem to have been quite successful, albeit expensive, in the American Southwest. In short, worn-out car and small truck tires are used structurally as below-grade walls. The tires are "rammed" with earth, so powerfully that they actually inflate somewhat. I have read that a rammed tire can weigh up to 350 pounds. By all accounts, the physical act of hand-ramming the tires, often with sledgehammers or other tools made for the purpose, is one heckuva lotta work. There is a very real danger of developing tennis elbow by the time you've finished the third tire. Finally, there is the issue of burying hundreds of tires in the ground, although, in fairness,

Earthship advocates can come right back and criticize me for the stuff I bury in the ground. The fact is that many people love these homes, including the actor Dennis Weaver, who had a very large one built for him. It looks great in the video. Readers interested in this style of earth-sheltered home can find more information at their website, <www.earthship.org.>.

MIKE OEHLER'S "PSP" METHOD

PSP stands for Post, Shoring, and Polyethylene. Friend Mike has been building with this method for many years, and has convinced others to do the same, particularly those on a very low budget. When it comes to economy, I have to admit that Mike outflanks me. The posts are new or recycled posts, their bottoms buried in the ground. They appear frequently around the perimeter of the building and are left exposed on the interior. The "shoring" is planking, often recycled. Inch-thick planking can be used if the posts are quite close together, two-by planking if the posts are further apart. I'm not going to give specifications on this, believing that "A little knowledge is a dangerous thing." Read Mike's book. The third component is good old-fashioned 6-mil polyethylene, also called polythene or Visquene after a commercial brand. This 6-mil sheet is the waterproofing barrier. It goes between the shoring and the backfill.

Mike's first house cost $50 using this system and he got three years out of it before a bear moved in. Later, after the bear had moved on, the original $50 structure became a part of Mike's $500 house, where he still lives. If you are interested in this method, or just want to read a radical and entertaining book, get Mike's *The $50 and Up Underground House Book*. (See Bibliography.)

And so this leaves us with my favorite method, the one I think is most appropriate for inexperienced owner-builders:

SURFACE-BONDED BLOCKS

This little-known construction method, developed by the US Army Corps of Engineers, has been around for over 40 years. In short, it consists of applying – or "bonding" – a strong tensile membrane to each side of a wall of dry-stacked concrete blocks. The wall is many times stronger on tension than a conventionally mortared wall with its weak mortar bond. The tensile strength of surface-bonding cement comes from the millions of ½-inch glass fibers that permeate the mix. Surface bonding (SB) cement is normally applied ⅛-inch-thick to each side of a dry-stacked block wall.

Twenty-five years ago, I wrote: "A conventionally mortared and a surface-bonded block wall cost about the same when all factors are considered. The mortared block wall requires more skill and more labor than does the surface-bonded wall, yet the surface-bonded wall is much stronger against lateral pressures. For a home with straight walls, use 12-inch-thick concrete blocks in conjunction with surface bonding. This method is strong, moderate in cost, easily learnable, and relatively fast, and it supplies an excellent base for waterproofing outside and painting inside." I still feel the same way.

LOG END CAVE: THE WALLS

In Chapter 3, I promised to explain the rather odd wall measurements that we wind up with when using the surface-bonded method of block construction. The reason for strange numbers, both in height of wall and length, is that blocks are all 7⅝ inches high and 15⅝ inches long. With standard ⅜-inch mortar joints, the modular size of the blocks can be thought of as a nice round 8-by-16-inches. But, when the mortar joint is eliminated, as we do with surface-bonding, multiple courses of blocks – and lengths of several blocks strung together – yield inconvenient numbers. This is why we use a chart like Table 3 to do the math for us. Multiplying by 7⅝-inch or 15⅝-inch is difficult, even with a calculator.

For example, if we have a wall made of 12 courses of dry-stacked blocks, it will be 7- feet 7½-inches high. And a wall of 24 blocks, laid tight end-to-end will be 31-feet 3-inches long. (12 blocks is 15-feet 7½-inches; we double that number for 24 blocks.) In the case of Log End Cave, each course is completed with a 12" block turned the other way, so I add a foot to the number used for the whole number of lengthwise blocks, each course. That's how we get the sidewall dimensions in Figure 1.10.

Table 3
Dimensions of Walls Constructed with Surface-Bonded Concrete Blocks

NUMBER OF BLOCKS	LENGTH OF WALL[1]	NUMBER OF COURSES	HEIGHT OF WALL[2]
1	1'3⅝"	1	7⅝"
2	2'7¼"	2	1'3¼"
3	3'10⅞"	3	1'10⅞"
4	5'2½"	4	2'6½"
5	6'6⅛"	5	3'2⅛"
6	7'9¾"	6	3'9¾"
7	9'1⅜"	7	4'5⅜"
8	10'5"	8	5'1"
9	11'8⅝"	9	5'8⅝"
10	13'0¼"	10	6'4¼"
11	14'3⅞"	11	6'11⅞"
12	15'7½"	12	7'7½"
13	16'11⅛"	13	8'3⅛"
14	18'2¾"	14	8'10¾"
15	19'6⅜"	15	9'6⅜"

Note: blocks are actually 15⅜" long by 7⅜" high.

[1] Add one-quarter inch (0.25") for each 10 feet of wall to allow for nonuniformity in size of blocks.

[2] Make a trial stacking of blocks to determine the actual height of wall or wall opening before beginning construction.

The first course of blocks had already been mortared to the foundation, as mentioned in Chapter 4. Mortaring the first course of blocks to the foundation is correct practice with a surface-bonded wall. The purpose is not so much to "glue" the blocks down to the foundation; rather, this is the last best opportunity you have to get the top of the block course as level as possible, in both directions: along the length of the wall, and across its width. The only difference from

*Fig. 5.3:
The author lays
the first blocks,
in mortar, at
Log End Cave,
keeping blocks
tight against
each other.*

standard mortared block practice is that we don't leave a ⅜-inch mortar joint between adjacent blocks; we lay them down pretty tight, one to the other. If you don't feel confident laying your own first course, you could get an experienced block mason to do it, but, if he or she is unfamiliar with surface bonding, impress upon the mason the need for getting good level in each direction laying the blocks tight to each other along the length of the wall. I am not a block mason, but I prefer to do this job myself. Yes, it takes me longer, but I get it the way I want it and I save money on labor costs.

For mortar, I like the old standby all-purpose recipe of three parts sand and one part masonry cement. Whenever you're stuck for a mortar mix, or even plaster, you won't go very far wrong with: 3 sand, 1 masonry cement. Alternatively, you can purchase pre-mixed "mortar mix" in a cubic-foot bag. Just add water. In either case, mix in enough water to get a moderately stiff mortar. You don't want it so wet that the heavy block sinks into it. You want to be able to go a little thicker, if need be in low spots, and still support the block. Dampening the underside of the block will get a better "paste bond" to the block. I also dampen the foundation very slightly, and for the same reason.

Remember that the first course is keyed to the foundation by one of the methods described early in Chapter 3, under "Resisting lateral load on the wall." If you choose the keying method of Figure 3.2 – my recommendation – you can pack some concrete into the bottom half of the block cores. It is probably safest to wait until the next day to do this, so that the mortar joint has set. You can mix your own concrete in a wheelbarrow using a mix of 3 parts pea gravel, 2 parts sand and 1 part Portland cement, or you can buy ready-mixed bags of concrete mix (not mortar mix) made by Sakrete™, Quikcrete™ and many others. Mix the concrete fairly stiff. Pack and vibrate it into the block cores with a stick.

Your first course is now in place and keyed to the foundation to help resist lateral pressure from the backfill. Subsequent courses are stacked "dry," that is, without mortar.

I chose 12-inch-wide blocks for the Cave walls for stability and to eliminate the need for pesky pilasters which would be necessary with the more conventional 8-inch-wide blocks. See Figure 5.4 and Figure 5.5. Avoid 10-inch blocks like the plague. They are the "blocks from Hell" requiring lots of fitting and cutting, often on each side of each course.

Actually, with 12-inch block walls, there is a very real practical advantage in "turning" a course at the end of a wall run with a 12-inch corner block, as seen in Figure 5.6. You can stagger the vertical "joints" without cutting blocks, as per the left side of the diagram. On the right, we start with a whole number of blocks on the first course, and then we are stuck with cutting a block on the next course. Using this strategy at the Cave, all we had to do, once per course, was to cut 4 inches off of a single 4-by-8-by-16 solid concrete block, a common item. Do this by marking the block with a pencil and scoring the pencil line with a hammer and cold chisel. Keep working your way around the block, hitting a little harder each time, until the block breaks off. If you decide on 8-inch-wide blocks, and pilasters, you will use a half-lap on the second course. Eight- by-eight-by-eight "half" blocks are a commonly available item, and are used at the ends of courses and around windows and doors.

In some populated areas, you may find "full modular blocks" made deliberately for surface bonding. Slightly over-sized blocks are passed through a surface grinder to give uniform blocks that actually measure 8-by-16-inches by whatever thickness they might be. I have heard of these

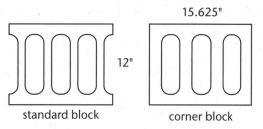

standard block   corner block

Fig. 5.4: Standard 12" blocks have half-cores at their ends, while corner blocks are square-ended.

blocks, but the closest I have seen to them are one-foot-by-one-foot-by-two-foot lightweight blocks made for insulated above grade construction. It doesn't hurt to ask.

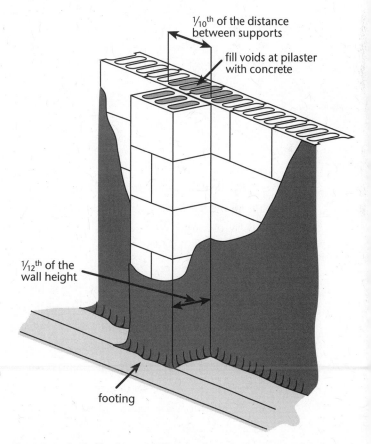

¹⁄₁₀th of the distance between supports

fill voids at pilaster with concrete

¹⁄₁₂th of the wall height

footing

Fig. 5.5: A pilaster like this can stiffen a wall or provide extra support for a girder. Pilasters are most useful when using 8" concrete blocks.

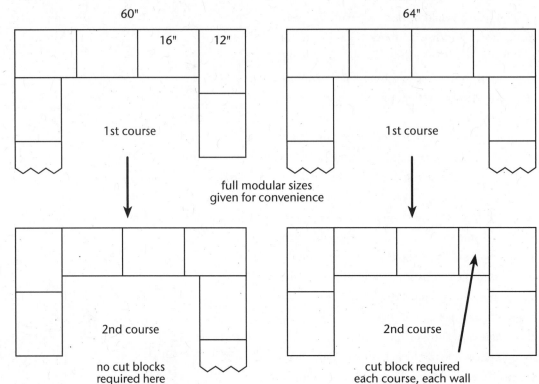

Fig. 5.6:
By tweaking
plans at the
design stage,
corners can be
made without
having to cut
blocks.

Blocks should be cleaned of burrs and rough bits of concrete by scraping them across the top of a pallet of blocks, or by using the flat edge of a piece of broken block as an abrasion tool.

Build the corners up by stacking blocks dry, and check for plumb and level with a four-foot level. Use only corner blocks – the non-scalloped ones – in the corners. After the four corners are each built up three or four courses, as in Figure 5.7, stretch a nylon mason's line from corner to corner and place the other intermediary blocks up to the line, always keeping blocks tight against each other.

Sometimes, blocks will tend to wobble a little on the previous course. Thin metal shims can be used to take the wobble out, but use your level to find out which of the two "loose" corners would be better off with the shim. I've cut useful shims from aluminum flashing and offset printing plates from a print shop. Use your tight mason's line and your eyeball to keep courses straight and level.

Did I tell you that ordinary 12-inch concrete blocks can weigh 65 pounds? And corner blocks five pounds more? Well, that's the bad news. The good news is that, except for the first course, you don't have to lift them over a mortar joint and set them down carefully. You will be impressed at how strong a dry-stacked wall of 12-inch blocks is even before they are surface-bonded. You can be pretty rough with heaving subsequent blocks onto the wall, within reason. With a building like Log End Cave, which was only 11 full courses high, you can actually stack the whole building before applying any SB cement.

Applying the Surface Bonding Cement

Ask at your local masonry supply or building supply store for surface-bonding cement, abbreviated SB in this book. Some people still call it "block bond," a holdover from the days when there was a SB cement by a similar name. It generally comes in 50-pound bags, which will cover 50 square feet at ⅛-inch of thickness. The main ingredients in SB cement are Portland cement and millions of ½-inch glass fibers. Conproco, a manufacturer in New Hampshire, uses Type AR (alkali resistant) fibers, the only kind acceptable, for example, in Massachusetts. It is likely that any SB cement sold in your area will meet local code, but it never hurts to check. There are other trace ingredients, such as calcium stearate, used as a waterproofing. Several SB cement manufacturers are listed in Appendix B: Sources. They are generally quite helpful about supplying additional information about their products, explaining hot- and cold-weather applications, for example, and giving additional tips on mixing and applying the cement.

Soak the wall thoroughly before commencing the application of the SB cement, to prevent the rapid transfer of the cement's precious moisture to the blocks. The wall should be saturated, but not dripping.

Mix the SB cement in a wheelbarrow, according to the instructions on the bag. The cement should be quite wet, like a very thick cream, but not so wet that it falls off the wall during application. When you've got it right, you get an extremely good bond to the dampened block wall, and it is easy to lay on a smooth consistent coating.

Use a flat plasterer's trowel to apply the SB cement, as seen in Figure 5.8. You can use a regular

Fig. 5.7: After the first course of blocks is in place and level, build the surface-bonded wall by stacking blocks three or four courses high at each corner. Then, fill in between corners by working to a mason's line stretched from corner to corner.

Fig. 5.8: The author presses the surface-bonding cement firmly into the dampened wall with a plasterer's trowel.

a garage, say – it would be very important to put the full ⅛-inch coating on both sides of the wall, creating a kind of stress-skinned panel. After all, you don't know which way the wall might want to fall over. Underground, however, we know which way the wall wants to fall: inward. (The wall can't fall out because of the tons of earth loading it.) So make sure you get that honest ⅛-inch on the interior surface. On the exterior side, a good honest ¹⁄₁₆-inch will suffice. I checked with one of the SB cement engineers on this, and I only mention it because SB can be quite expensive, about $16 per bag in June of 2005.

Using a hose or garden sprayer, mist the SB cement frequently after it sets a bit, at least twice during the first 24 hours, more if you are suffering from rapid drying conditions of sun and wind. Keeping the wall damp reduces the incidence of crazing and shrinkage cracks. By paying attention to the travel of the sun, you can almost always work in the shade. Do the west sides of the walls in the morning, for example, and the east sides in the afternoon. North faces can be done safely most anytime, and south faces are best done last thing in the day.

Additional Strength for Long Straight Walls

Internal shear walls. A long straight wall is subject to bending pressures under the earth backfill. Like the underside of a beam, the inner surface of a straight earth-loaded wall is on tension. Surface bonding cement, of course, lends tensile strength

pointed trowel to load up the flat trowel with SB cement. Then, beginning at the bottom of the wall, apply the mix with firm pressure on the trowel, pushing the cement upward so that a uniform covering is obtained. You can even out the plastered area and spread excess SB cement with long light strokes, always holding the trowel at about a 5-degree angle to the wall surface. Avoid over-troweling, which can cause crazing and cracking of the surface. With a little practice, you'll soon be laying a pretty consistent ⅛-inch coating onto the wall. Until you feel confident, you can check the depth with the corner of your trowel.

A word about the ⅛-inch inch recommendation: If you were building a wall above grade –

to that surface, but, alone, might be inadequate on extreme spans. The best, easiest, and cheapest way to address these bending pressures is at the design stage. Including an internal shear wall, halfway along the long wall span, effectively cuts that span in half, making it four times stronger against bending pressure from without. The shear wall, like an internal brace, can be concrete block or wooden, but, in either case, its base must be well-fastened to the floor. Figure 5.9 shows how we used internal shear walls as braces to shorten spans at Log End Cave. Obviously, compromises in the floor plan may be necessary to accommodate these intersecting internal walls.

Bond beams. Another way of adding additional tensile strength to a wall is by the use of bond-beam courses of block. Special bond-beam blocks are made with either one or two troughs running lengthwise along the block. These blocks are laid end-to-end around the building, making a continuous trough (or double trough) around the building. Before proceeding to another course, this long trough is filled with concrete and continuous ½-inch or ⅝-inch rebar, lapping rebar by 40 times its diameter (20-inch lap for ½-inch rebar, 25-inch lap for ⅝-inch rebar). Effectively, this casts a continuous reinforced concrete beam around the building. The piece of rebar on the tension side of the wall is the one that works towards resisting the lateral thrust of the earth backfill. The most effective courses for bond beams would be the top course, typically the 12th course, although this can vary with wall height,

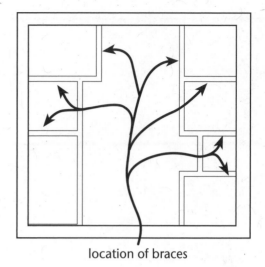

location of braces

Fig. 5.9: Internal braces, strategically placed in the floor plan, greatly increase the wall's ability to resist lateral pressure from the earth berm.

and a course about half way up the wall, the 6th or 7th course. You may wish to look ahead to Figure 5.11 to see a sectional drawing showing a bond beam course in place.

Vertical reinforcing. At Log End Cave, we created a number of vertical solid concrete "pillars," hidden in the wall. We placed them at the strategic locations where the load from the three major girders would bear down on the wall – roughly 9-foot centers – and three more on the east and west walls, on about 7-foot 6-inch centers. Make these pillars by slushing two or three consecutive vertical block core cavities with concrete and vertically installed rebar. With half-lapped 8-inch blocks, filling two consecutive cores guarantees that adjacent blocks are always "stitched" together by the rebar. With the type of 12-inch blocks we used, three adjacent cores had to be filled to guarantee the stitching feature.

Fig. 5.10: Earthwood: the first course of 8" corner blocks is set in mortar.

The prudent course of action is to have your wall plans checked by a structural engineer, who can advise you on the need and placement of bond-beam courses and filling cores for bearing pillars.

Surface Bonding at Earthwood

At Earthwood, the intent had been to build 16-inch cordwood walls around the building's curved earth-sheltered northern hemisphere, hardwood log-ends in particular, for greater thermal mass. When the hardwood swelled, causing a loss of plumb in the wall, we returned to the tried and proven SB walls, but with a "twist." We still wanted 16-inch-thick walls (to support cordwood masonry on the second story), so we decided to lay 8-inch corner blocks (8-by-8-by-16-inch) transversely in the wall to achieve the required thickness. We needed corner blocks exclusively, so that we would not be showing the scalloped ends of regular blocks, inside and out.

My first estimate was that we'd need 1,700 blocks for the entire job. Initially, I ordered 1,100 corner blocks for the lower story. My supplier at the block plant was not happy about this, as his normal production run is about 30 percent corner blocks and 70 percent standard blocks, but he went along in preference to losing the sale. I also added a hundred 4-inch solid blocks to the order, as I wanted a solid 4-inch course just before the first-floor joists were to be installed.

As at the Cave, the first course was laid in a mortar joint, as seen in Figure 5.10. Note, in the picture, that there is a groove on one end of each corner block, used for locking window and door frames in normal horizontal block construction. As we were laying the blocks transversely, I decided to leave the useless groove to the outside every time, in order to have a smoother interior wall surface.

The inner end of the blocks are laid tight to each other, but, because the outer circumference of the 16-inch-thick wall is greater than the inner circumference, there will be a gap of about 0.55-inch (9/16-inch) between blocks on the outside. Later on, prior to surface-bonding, I filled both this gap and the irritating grooves with my all-purpose mix of 3 parts sand and 1 part masonry cement.

Again, shims are used to take any wobble out of the blocks. Use a level to determine where a shim is best placed.

As we dry-stacked the blocks, we filled the cores with sand to increase the thermal mass of the below grade wall. Filling these cores with

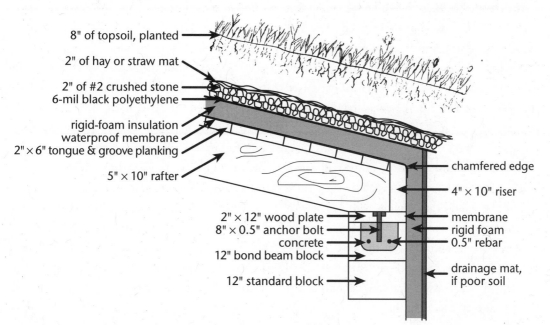

8" of topsoil, planted
2" of hay or straw mat
2" of #2 crushed stone
6-mil black polyethylene
rigid-foam insulation
waterproof membrane
2" × 6" tongue & groove planking
5" × 10" rafter

chamfered edge
4" × 10" riser

2" × 12" wood plate
8" × 0.5" anchor bolt
concrete
12" bond beam block
12" standard block

membrane
rigid foam
0.5" rebar

drainage mat,
if poor soil

Fig. 5.11: Detail of the fully bermed east (or west) wall, from 40 × 40' Log End Cave plans.

concrete is unnecessary for strength, and would be a lot more difficult, time-consuming, and expensive. Always keep in mind that the inner surface of a curved wall is on compression. As the wall is loaded, the lines of thrust are lateralized around the wall, finally stopping at the buttresses. The relatively small load of backfilling, even two stories, is not going to exceed the crushing strength of concrete blocks.

Filling the cores with insulation would be a huge mistake, negating the greatest part of the block wall's mass. The insulation will be correctly placed to the exterior after the waterproofing membrane goes on, as discussed in Chapter 7.

LOG END CAVE: FINISHING THE TOP OF THE BLOCK WALL

At the Cave, we made the transition from block wall to timber framing with the aid of a two-by-twelve wooden sill plate anchor-bolted to the top course.

Anchor bolts are typically a half-inch in diameter and 8 to 12 inches long. An inch or two of one end is bent to a right angle, an "ell," to grab the concrete. The other, protruding, end is threaded to receive washers and nuts. We employed two different methods of setting anchor bolts into the last course of blocks. In locations where the two or more adjacent block courses were totally filled with concrete and vertical rebar, the anchor bolts were set in the concrete while it was still workable. If an anchor bolt is required at a place not already slushed with concrete, we would stuff a large wad of heavy paper about ten inches into the cores and fill that 10-inch void with fresh concrete. Sakrete™ or equal pre-mixed concrete is perfect for the job, and part of a cement bag is great as the paper wad, so don't be too quick to trash them.

If you have finished the top of the wall with a bond-beam course, you can install the anchors as you fill the bond-beam troughs with concrete. It is recommended to hook the anchor bolt's ell under the horizontal rebar, in this case, for even greater strength, as seen in Figure 5.11.

A good way to assure that the anchor bolts are plumb and at the right height is to make a template out of the same material that you'll be using as the sill plate, such as a piece of two-by-twelve, in the Cave's case. Drill a hole vertically through the template piece, just a wee bit larger than the diameter of the bolt – a $\frac{9}{16}$-inch hole for a $\frac{1}{2}$-inch anchor, for example.

The location of the anchor bolts around the top of the wall is a function of the lengths of your sill plate material. Two bolts are sufficient for short lengths, up to 8 feet. Keep the anchors a foot from each end, in that case. For sills longer than eight feet, you'll want three anchors or more. An anchor bolt every four or five feet is enough.

Rough cut 2-by-12s covered our 12-inch-wide block walls perfectly on the north wall and the short south wall. The east and west walls had four-by-eight plates at the original Cave, seen in Figures 5.12 and 5.13, but I now prefer the 2-by-12 plates all around, as per Figure 5.11.

Let the concrete supporting the anchor bolts set for at least two days before attaching the plates. Working prematurely could cause the bolt and the concrete to lift, and then you are back to square one.

To get hole positions on the sill plate to match the anchor bolt locations, simply place the sill plate over the bolts and whack the timber once with a heavy hammer, as per Figure 5.12. As in life, "You only get one chance to make a good first impression." A second helper is very useful to hold the plank in position until an impression is made for all bolt locations. Next, turn the plank over and, where you see the impressions, drill holes straight through the plank, holes of the same diameter as the bolts. Use a brace and bit, as per Figure 5.13, or an electric drill fitted with a spade bit.

Fig. 5.12: Wait for the anchor bolts to set in the concrete plugs in the block cores. Then, locate the holes by laying the plate on the anchor bolts and whacking once sharply with a heavy hammer.

Usually, bolts protruding even a half-inch above the plates might be in the way of the next phase of construction, setting windows, for example, or continuing with a framed wall. For this reason, we countersunk the washers and nuts below the top surface of the sill plates. The countersink depression should be about ⅝-inch deep and a little bit larger than the flat washer that goes under the nut. Most ½-inch washers are a little over an inch in diameter, so a 1¼-inch spade bit would be an appropriate tool for creating a good countersink. Wrap a piece of tape around the shaft so that you don't exceed the ⅝-inch depth. Or simply check the depth frequently with a rule.

Prior to fastening the sill plates, lay out a course of ¼-inch flexible polyethylene sill-sealer, made for the purpose. Often blue in color, and sometimes going by the trade name Sill Seal®, these little rolls of material give a good protective seal between wood and concrete. Finally, we insert a good ½-inch flat washer, and fasten the plates down with a half-inch nut. A socket wrench works best.

SIMULTANEOUS EVENTS AT EARTHWOOD

We divide a book into chapters for convenience, discussing one component of construction at a time. In reality, at Earthwood, three different supporting components moved forward simultaneously: the external walls (both surface-bonded blocks and cordwood), a central stone

mass (doubling as a masonry stove), and an octagonal post-and-beam framework halfway between. All three structural elements needed to be in place to support the radial floor-joist system, so that we could proceed on the second story.

Cordwood masonry is an option for external walls, one near and dear to my heart, but it is too specialized and complex a subject to include in this book. Readers who fancy cordwood masonry for above-grade walls are referred to my *Cordwood Building: The State of the Art*, listed in the Bibliography.

As for the masonry stove at the center of the building, its construction is also too specialized to include in this volume, so, again, the reader is referred to the Bibliography. And there is the added issue of code compliance. Massachusetts, for example (and probably other states) define the masonry stove as a chimney, and there are code

Fig. 5.13: Turn the plate over and drill holes through it of the same diameter as the bolts. See text for countersink procedure.

A Popular Misconception

Many of our students have great difficulty letting go of the 3/8-inch mortar joints and suggest that the wall would be even stronger if it was mortared up first, and then surface-bonded. This is not so. The top of Figure 5.14 shows the joint between blocks in a regularly mortared wall, while the bottom shows the joint between two dry-stacked blocks. The little lines running in various directions show some randomly scattered half-inch glass fibers scattered near the block joint, in either case, much as they would be inside a layer of SB cement.

block

mortar

block

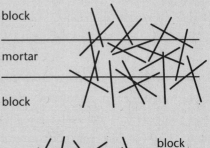

Note that on the mortared wall, very few of the fibers actually span from block to block. Most join a strong block to a weak mortar joint. Below, virtually every fiber in the vicinity of the joint actually spans from block to block. Suspended in the strong

block

block

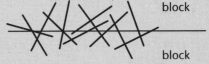

Fig. 5.14: The join between mortared and non-mortared blocks.

Portland cement medium, this creates a strong tensile membrane, always bonding block to block, not block to weak mortar joint.

regulations to prevent contact between chimneys and wooden members. We, too, were concerned about the safety of using a massive masonry stove as a part of the support structure. There is a quality of wood aging called *pyrolysis*, literally "chemical degradation by heat." For our application, pyrolysis can cause a lowering of the combustion temperature of wood. The Earthwood plan, as we built it, has the inner end of the floor joists bearing on a stone shelf just five feet from the masonry stove firebox. My theory was that the great mass of the lower portion of the stove, roughly 15 tons of stone and other masonry materials, would absorb enough heat to keep pyrolysis out of the equation. I suppose I was taking a chance, but stayed the course, nonetheless.

Months later, we were able to check the surface temperature of the masonry mass at the point where the rafters bear on the stone shelf. Even with double-firing of the stove, getting it as hot as we would ever get it, we could not quite reach 100 degrees Fahrenheit (38 degrees Celsius) on that shelf, a temperature equal to a hot summer's day. No crisping of the floor joists has ever occurred. Having said all that, and having satisfied myself that my family is safe, it must be stated that with any form of alternative building methods, the onus is on the builder to convince the code enforcement officer that his or her new methodology meets or exceeds the intent of the code.

If the reader is unable to convince his code enforcement officer of the suitability of the Earthwood plan as designed and built, an alternative is to build a small eight-sided octagonal post-and-beam frame close to but at least two inches away from the mass. The tops of the eight posts are connected by eight short girts, which, in turn, support the inner ends of the radial floor joists.

As for timber framing, joists and rafters, well, that is the subject of the very next chapter.

Chapter 6

Timber Framing

Heavy timber framing is also known as post-and-beam construction. In the past few years, there has been a resurgence in traditional finely crafted wooden-jointed timber framing. While I have the highest admiration for the craftspeople in this trade, I have never had the skill, time, or patience to do this caliber of work myself; and, truth be told, neither do most contractors. Therefore, the methods I accent at Earthwood, and in this book, are the common methods that most people actually do: joining timbers with readily available mechanical fasteners.

The natural bedfellow to post-and-beam framing is plank-and-beam roofing, the best choice, in my view, if the goal is a low-cost, beautiful, high-quality earth roof. I am not alone in deriving a sense of security from seeing the heavy timbers next to me and over my head. We know why we are safe; the evidence is all around us.

It is good to clarify some terms. *Beam* is a much over-used term, and one that is often misused. For our purposes, the beam of post-and-beam can be more definitively called a *girt*

(around the perimeter of a structure) or a *girder*, flying somewhere through the interior of a building. This girt or girder, in turn, can support the beam of a plank-and-beam structure, which might be a floor joist or a rafter.

About Posts and Planks

The strong components of post-and-beam and plank-and-beam systems are the posts and the planks. Posts – also called columns – are very strong. An eight-foot high eight-by-eight post, for example, even one with a relatively low compression rating of 1,150 to 1,400 inch-pounds per square inch (you don't have to understand the term) will support 63,000 pounds. At Earthwood, our fully saturated earth roof weighs a maximum of 207,750 pounds, 68 percent of which is carried by the external walls and the central mass. Not one of the eight-by-eight posts at Earthwood is called upon to support more than 12,500 pounds, about a fifth of its capacity. Even 6-by-6 posts would do the job, but I prefer 8-by-8s for ease of joining girders on top of them.

The planks are the other strong component. I used full-sized 2-by-6 planks at Log End Cave, planed on one side to make them uniformly 1¾-inch in thickness. Since then, I have come to prefer V-jointed two-by-six tongue-in-groove planking. This material is more expensive, but looks great overhead. The V-joint shows on the ceiling, while the top surface fits tightly together without a gap of any kind. I use spruce, white pine, or something oddly called "spruce-pine." With the tongue-in-groove stuff, or the thicker locally milled non-T-in-G material, the planking has the capacity to span over rafters set 4 feet on center and still carry loads greater than those of a reasonable – say eight-inch thick – earth roof with a 70-pound snow load. Long before we exceed the carrying capacity of the planking, we will run into engineering problems with the rafters.

DESIGN CONSIDERATIONS FOR RAFTERS

If you can plug numbers into formulas and solve for a single unknown, you can use Appendix C: Stress Load Calculations to check rafters and girders for both bending and shear strength. (Bending and shear are discussed on the next page.) And, while *Timber Framing for the Rest of Us* covers the subject matter of this chapter in more detail, you should, at the least, be aware that there are five different structural considerations that need to be integrated for effective design.

1. *Load.* As we saw in Chapter 1, a six-inch thick earth roof, fully saturated, and with a 70-pound snow load (as in Northern New York) will weigh about 150 pounds per square foot. This is more than twice the engineering load for non-earth-roofed buildings in our area.

2. *Kind of wood.* Every species and grade of wood has a different unit stress rating for shear and bending strength, so know your wood. There are pages of unit stress ratings for various species in engineering manuals.

3. *Frequency of rafters.* Simply stated, frequency refers to *how many* rafters you will be using. Will they be a foot apart (12 inches o.c.), two feet apart (24 inches o.c.), or some other spacing? All else being equal, doubling the frequency of rafters actually doubles the engineered strength of the structure.

4. *Dimensions of rafters.* Are you using 4-by-8, as at the Cave, 5-by-10 (Earthwood), or some other cross-section shape or dimensions? Bigger is stronger, but there is no point in over-building.

5. *Span.* Ah, span: the one that throws most designer-builders for a loop. What clear span are you trying to achieve? The key here is in knowing – and remembering – that the strength of a beam is inversely proportional to the

square of the span. This is critical. It is intuitive to think, for example, that a 12-foot span would have to be engineered 20 percent stronger than a 10-foot span, all else being equal, but this would be bad intuition. Ten squared is 100, but twelve squared is 144. The 12-foot span needs to be designed *44 percent* stronger!

You need to know four of the five variables described above to calculate the fifth. If you know the *kind of wood*, the *cross-sectional dimensions*, the *span*, and the *desired load*, for example, you could calculate how many rafters (the frequency) you need. Or, if you know the kind and grade of wood available, you can calculate what load a certain rafter system can support.

BENDING AND SHEAR

There is one more design consideration that designers of earth roofs, in particular, should be aware of, and that is the difference between bending and shear failure. Bending failure, the one we more commonly think of, occurs when a beam breaks near the center under an excessive vertical load. Rafters are usually much greater in their depth (d) dimension, than their breadth (b), in order to optimize bending strength. Bending strength – in addition to other considerations covered in the previous part – is predicated upon something called *section modulus*, a mathematical evaluation of cross-sectional shape. The formula for rectilinear beams is $S = bd^2/6$, where S is the

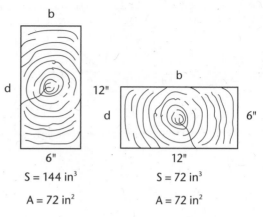

b

d   12"

6"

S = 144 in³

A = 72 in²

b

d   6"

12"

S = 72 in³

A = 72 in²

Fig. 6.1: Comparing the cross-sectional characteristics of beams installed in different ways.

section modulus, b is the breadth of the beam, d is the depth of the beam, and 6 is the appropriate constant for members with a rectilinear section. In our timber framing classes, I like to use the example of a 6-by-12 rafter (see Figure 6.1), because the math comes out so neat and tidy.

Note that the value of the depth of the rafter is squared, while the breadth is not. Section modulus gives a lot more credit for strength to depth than it does to breadth. The rafter shown on the left of Figure 6.1 has a section modulus of 144 inches cubed (not cubic inches!) when installed in the regular way, but only 72 inches cubed when laid up incorrectly, as on the right. This makes sense intuitively – correct intuition this time – because a plank lying flat and spanning from one support to another is springy compared to the comparative stiffness of one laid like a rafter.

Shear failure is the tendency of a beam to "shear off" right near a support, as per Figure 6.2,

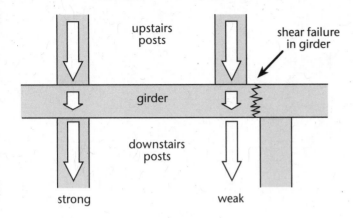

upstairs posts

shear failure in girder

girder

downstairs posts

strong

weak

Fig. 6.2:
The arrows
indicate lines of
thrust from the
roof.

below. A certain number of wood fibers have to "shear through," if you like, and this is the same whether we lay the member "correctly" or "incorrectly," because the cross-sectional area (A) of the member (in indication of the number of fibers) is the critical relationship, not section modulus.

With lightweight two-by stick framing, bending strength seems to come into play before shear strength, but, with heavy timber framing, the opposite is often the case. A little later, we'll discuss how shear came into play at Log End Cave, but, first, let's speak a little of timber procurement.

SOURCES OF TIMBERS

Recycled Timbers

At both the Cave and Earthwood, I made use of a lot of recycled barn beams, strong old timbers ranging from 8-inch-by-8-inch up to 10-inch-by-10-inch. Such timbers are particularly useful –

and beautiful – as posts. Posts are the real strong component of post-and-beam construction, and are usually way overbuilt. To give you an idea of the strength of posts, a 6-inch-by-6-inch first quality white or red pine post will support 32,000 pounds – 16 tons – at eight feet of height. Nine such internal posts, as seen on the 40-by-40-foot Log End Cave plans (see Figures 1.13, 1.14 and 1.15), would support 288,000 pounds, or 144 tons! Those nine posts actually carry the load of 900 square feet of the roof. (External walls carry the rest of the 40-by-40-foot plan.) The posts, therefore, support a roof structure of 320 pounds per square foot, way beyond our needs.

I actually used 8-by-8-inch barn beams at the original Cave, and even though they had occasional mortise joints in them and lacked a grade stamp, I've no doubt that they are at least as strong on compression as the freshly milled 6-by-6-inch pine. And I do reject, out of hand, any old beams that show rot, either from moisture or insects.

Where do you find recycled materials? Out in the country, right? Well, yes, that's where I have been finding them for the past 30 years. But, surprisingly, there is often even more demolition of old building occurring in cities, with lots of great opportunities for excellent salvaged timbers at reasonable price. City or country, the trick is to ask around. You need to "cultivate coincidences" by asking around. Once you get on the trail, you'll be amazed at what's out there.

Rough-cut Timbers from a Sawmill

Heavily forested rural areas have lots of local sawmills. Some use old circular saws and others have bandsaw mills, but the quality is quite good and the price is generally half of what you have to pay at the big national lumber suppliers. And your money stays in the county. If there is a drawback to the use of local rough-cut (full-sized) dimensional timber, it is that some states require that all structural timbers be grade-stamped for quality. New York was faced with such a threat when the new International Building Code came into play in 2003. Fortunately, the state code officials passed a variance that allowed the use of rough-sawn non-graded timber to be up to the discretion of local code enforcement officers. If that doesn't fly in your state, as it did not for Mark Powers, building in Michigan – his case study appears in Chapter 12 – then there is the option of hiring a professional grader to come in and grade the timbers you purchase, to meet code. This adds an expense, but it's usually still cheaper than buying graded heavy timbers at a national chain store.

Milling Your Own Timbers

You might hire a local sawyer to bring his portable Wood Mizer or similar bandsaw mill to mill your own trees into heavy timbers or other lumber. Or, you might choose to purchase one of the variety of chainsaw attachments that make it possible for you to mill your own with a good quality chainsaw.

All of these procurement suggestions are covered in detail in Chapter 3 of my book, *Timber Framing for the Rest of Us.*

TIMBER FRAMING FOR EARTH-SHELTERED HOUSING

With earth-sheltered housing, in particular, it is important to integrate the floor plan with the structural plan. A judiciously placed post-and-beam frame should be used to cut roofing spans down to something which is affordable. The location of posts and beams, in both structures, is shown in the plans in Chapter 1, particularly in Figures 1.10, 1.12, 1.13, and 1.15 through 1.17.

In each case, posts are integrated with internal walls wherever possible. In each home, there is a free-standing post which some might consider as "being in the way." We get used to them and take comfort that they are saving us a small fortune in the added structural cost of doubled spans, which would quadruple the structural requirements and, probably, cost.

At both homes, ordinary 2-by-4-framed interior walls meet the sides of the 8-by-8-inch posts, leaving the posts a little proud of both sides of the wall, a pleasing esthetic feature. Sometimes intersecting walls meet at a post.

Overhead, a similar strategy has internal walls rising up and meeting the underside of the four-by-eight rafters at Log End Cave, the 4-by-8 floor joists on the downstairs level at Earthwood, and the 5-by-10 rafters upstairs. These framed walls lend strength to the underside of girders, joists

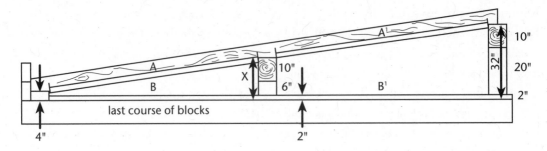

and rafters, and the members themselves, nicely exposed at the top of each wall, actually cut down on wall materials.

LOG END CAVE: TIMBER FRAMING

The structural plan was very simple at Log End Cave. The home was designed around the availability of three 30-foot 10-by-10 girders, which divided the home into four north-south bays. These girders, in turn, would support 4-by-8 rafters on 32-inch centers. The north and south external walls would support the ends of these large girders assisted by three internal posts in the case of the eastern and western girders, and two posts for the larger and stronger central girder.

To establish our 1.5:12 roof pitch, the plan called for a 20-inch mini-post between the top of the north wall plate and the underside of the 10-by-10 central girder, as seen in Figure 6.3. We called the concrete floor level and based the height of the two internal support posts on the total elevation of the girder off the floor at its north end, which was 95 inches to its underside. In an effort to key the posts to the girder, we cut the two interior barn beam posts at 96 inches, so

that we could recess them an inch into receiving notches cut into the underside of the girder. I have since changed my view on this, as that one-inch check into the girder actually decreases its shear strength over the posts by 10 percent. It is better to just toenail – or "toe-screw" – the top of the post into the underside of the girder. This also eliminates the chance of a measuring error. Putting a notch in the wrong place on a beautiful 10-by-10 girder spoils the day.

The south wall post would be in two pieces, so was a little trickier to calculate. The block wall on the south side of the home was very much shorter than the other three block walls – just four courses, capped off with a 2-by-12 wooden plate. The base post, the one standing on the wooden plate, had to be 48½ inches high, in order to frame one side of our 48-inch-high thermalpane windows. With fixed window units like these, you need to allow a quarter inch of air all around the glass, so the rough opening of an 84-by-48-inch unit is actually 84½-by-48½ inches. But, back to the post calcs: The header over the window would be a 4-by-10 laying on its side, the 4-inch dimension to resist against sag over the window, and the 10-inch dimension to support a cordwood wall above it at the same width. So, the short post over the 4-by-8 window lintels can be figured by subtraction, beginning with the 95 inches to the underside of the girder: 95 inches minus 30½ inches (the short block wall and its wooden plate, by measurement) minus 48½ inches (the base post between two of the large front windows) minus 4 inches (the window header), leaving a short "post" of 12 inches.

The reason that the central 12-inch-high "mini-post" on the south side is 8 inches less than the 20-inch post on the north side is that the top of the 4-by-10 header was 8 inches higher than the tops of the plate on the north side, done deliberately to give enough door and window room. The 4-by-10 window and door header is fastened above the 4-by-8 plates we used on the east and west wall. Don't worry about the actual measurements in our case. The point is to figure this all out accurately for the home you want to build. The structural components at each end of each girder need to add up to the same number: the height to the underside of the girder. This is the same figure as the height of any and all intermediary posts.

We stretched a nylon mason's line from the top of the north-wall post to the top of the south-wall post. This revealed that one of our two central posts was an inch higher than the underside of the girder, and the other was ⅞-inch higher. This worked well with our intended purpose of keying the posts into the girder (which

I wouldn't bother with again for reasons already given), and showed that our floor was within ⅛-inch of level.

INSTALLING POSTS ON CONCRETE

Old barn timbers, carefully examined for quality, make attractive posts. Their strength is manifested as a visual statement, adding power of spirit to the obvious structural benefit, the same as with heavy timbers overhead as rafters and girders. After removing any old nails from the timber, you will need to square the end that is to stand on the floor. I use a chainsaw, although the job can also be accomplished by going all around the timber with a circular saw, following pencil lines established by a framing square. With an 8-by-8 post, there will be a small (roughly 3-inch-square) section of uncut wood, which can be finished off with a handsaw. With a chainsaw, cut into the timber corner at a 45-degree angle to the two adjacent surfaces, each marked with a pencil line. Cut simultaneously along both pencil lines and let the saw continue straight on through. If your chain is new or properly sharpened, *and* you keep a steady eye and hand, you should have a good square cut.

Try the post by standing it up on the concrete. If it is pretty close to plumb, you've got it. If not, mark it and do it again, making the post two inches shorter. (Make sure you've got the extra length!) Use a wood scraper, rasping tool, or circular sander to remove a bit of extra wood that might be spoiling the cut. When the post stands

Fig. 6.4:
We temporarily
braced the
support posts
for the main
girders.

observed deterioration at the bottom of any of our posts. I *have* seen deterioration on posts where this important step was omitted. I have had good success with cut squares of 240-pound asphalt roofing shingles, as well as W.R. Grace Bituthene® Waterproofing Membrane. The thick asphalt shingles can actually help steady the post. Protrusions on the post base cause a point load on the shingle, compressing the material and providing stability.

To lessen the likelihood of injury from falling posts, we waited until all eight posts at the Cave were cut, squared, and tried for plumb, before setting them up permanently. One at a time, we stood them on their damp-proof squares of asphalt shingles, and made them plumb and rigid with a rough framework of 2-by-4s, seen in Figure 6.4, at left. We did not fasten the bracing to the floor, nor did we pin the posts to the concrete. The tons of weight on these posts, and the support from intersecting internal walls later on, assure the posts will not move.

If the reader (or a code official) is squeamish about leaving posts unfastened at the floor, the best way to pin the posts to the floor is with leaded expansion shield and matching lag screws. Drill a hole into the concrete with the size of drill embossed on the side of the expansion shield, ⅝" D for example. Drill the hole ⅛-inch deeper than the length of the expansion shield. Blow out the concrete dust with a straw and tap the expansion shield into the hole until its top surface is flush with the floor. With a wrench, install the

nicely by itself, take it down and measure and mark the other end. You only get one try at the top end, so be careful. On the plus side, if it is a little off, the error can always be made up with wooden shingles.

There needs to be a "damp-proof course" between the floor and the bottom of the post. We have always installed such a course – a code requirement, by the way – and have never

appropriate lag screw (again, this is embossed on the expansion shield, as $7/16"$ S), selecting a length of screw that leaves 1½-inch extended above the floor, as a pin. Cut the hex head off with a hacksaw and you've got a nice positioning pin exactly where you want it. Press your damp-proof course over the pin and into position on the floor. Set the post where it belongs over the pin. Two adjacent corner blocks on the floor serve as a good positioning guide for getting this just right. One person holds the post in position while another, on a stepladder, whacks the top of the post and makes an impression. Take the post down, and drill the appropriate hole into the underside of the post with an electric drill. Re-erect the post on its pin. The pin will stop the post from wandering across the floor during construction.

The pinning technique described above is particularly valuable for setting heavy timber doorframes in the external wall. The use of two pins on each timber will stop the timber from rotating, very important with doorframes.

At the Cave, it was necessary to install the long window and door header, before the short (12-inch-high) "post" could be installed. Our lintel was made from two 17-foot-long 4-by-10 timbers cut at a local sawmill, and joined over the central post on the south wall. These pieces were pretty green and very heavy, and it took four of us to maneuver them into place. The southern ends of the other two 30-foot 10-by-10 girders would bear right on this lintel, right over the points where the 48½-inch vertical posts

separated the windows and the door. These posts are made from pieces of 8-by-8 barn timbers. Look again at Figure 1.11 and pictures in the color section. Remember, the three girders and the three windows that we'd scored influenced the structural plan in a big way, saving us a lot of money. For someone to match what we did would be a very much more expensive proposition. Design around bargains.

LOG END CAVE: PLANK-AND-BEAM ROOFING

The word "beam," this time, refers to the 4-by-8 hemlock rafters that I had custom-milled at a local sawmill. I chose 4-by-8s because they are stout enough not to twist under load. The full-sized 4-inch base is wide enough to sit firmly on the girders and sill plates, assisted only by toenails. Two-by rafters, on the other hand, would require strong blocking between rafters, to prevent twisting and to stop the old domino effect of rafters falling on their sides. The 8-inch dimension provides sufficient bending strength against the planned roof load. However, I didn't realize, when I ordered my material, that hemlock is not strong on shear. Theoretically, the original Log End Cave, as designed, and using hemlock rafters, is a little under the money with regard to shear strength.

There is another strange quality of shear that came into play, illustrated in Figure 6.6. The lower part of the illustration shows the situation at the Cave: a longer rafter supported halfway

Giant Girders at Log End Cave

Our three 30-foot "ten-by-tens" were stacked about a quarter mile away from the site. We installed them during a time when my nephew, Steve Roy, and his friend, Bruce Mayer, were visiting. Steve, Bruce and I took very careful measurements of the post locations for our little notches and transposed the measurements onto the timbers themselves. I'd selected the best and largest timber for the center girder, as it would be supported by just two internal posts between the north and south support walls. We labeled the other two "East" and "West."

We squared the ends of the timbers. The total length of the center support beam is 30-feet 3-inches, allowing its end to extend four inches beyond the support posts, a nice visual detail. The west beam was squared at a final length of 29-feet 11-inches, which gave us a two-inch reveal at each end. The most we could get out of the east girder was 29-feet 8-inches. Later, we added a three-inch piece from another beam for esthetics, making it the same length as the west beam. This did not compromise strength, as there was still full bearing on the north and south support walls. In all the years, no one ever noticed the piece added on to one end.

I recruited some able bodies while the boys finished up the post notches on the underside of the beams. Have I told you I wouldn't do this again? Again, it was a hectic afternoon of making last-minute preparations. By late afternoon, I'd gathered four additional strong bodies to supplement our crew. We backed my little Toyota pickup truck to the western beam, the "lightest" of the three, thinking that if we couldn't shift that one, we'd have no chance with the 600-pound center girder. With Jaki looking after baby Rohan, I was left with the easiest job, driving the truck. Six men lifted the north end of the beam while I drove the truck's bed under it. We tucked the timber right up to the back of the cab, so that the truck could take as much weight as possible. Then, paired at each end of three strong ash poles, the three two-man teams lifted the south end off the ground and their team leader told me to "Go!" Like a great crawling insect, we started up the hill road.

along by a strong girder. This configuration actually improves the bending strength on each of the two rafter spans. It also decreases "deflection" of the rafter, also known as "sag." However, shear

Halfway up the hill portion, I was chastised by six screaming banshees, in no uncertain terms, to slow down.

We deliberately overshot our driveway in order to back the beam to the site, still 100 yards away. We backed the girder so that almost half of it was cantilevered over the edge of the north wall, and then manhandled it from below into position on the posts, as seen in Figure 6.5. The beam's notches locked perfectly over the posts, to our great delight.

The team elected to tackle the heavier center beam next, reasoning that they might not be able to handle it after expending energy on another 500-pounder first. And the central girder had to be raised almost two feet higher than its fellows to the east and west, another challenge.

Fig. 6.5: We place the center girder on its posts, Log End Cave.

As with the first girder, we worked it halfway out into the building, and then put our tallest and strongest man, Ron Light, below the beam, catwalking with Bruce along some scaffolding. Like a power lifter on steroids, Ron heaved the south end up to where Paul and I could get our ash pole branch under it. Then, with everyone lifting together, we got it into place, beautifully positioned on the notches, except that one of the posts was too long. I told the crew that the beam would have to come off. "Maybe the post will compress under the load," said one of the tired grunts, hopefully. We were able to move the beam to one side, supported on a temporary post, while I removed a little off the top of the tall post with my chainsaw. With a chisel, Steve altered the check in the beam a wee bit. The second try was a success.

The last girder, by comparison, was a piece of cake. Easy for me to say … I was still driving the truck.

strength over the center support girder is actually decreased by the use of a single long rafter, as opposed to two shorter rafters meeting each other over the girder.

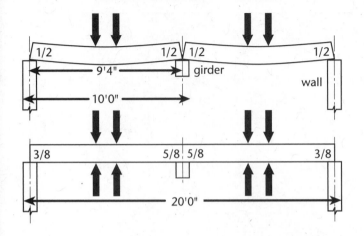

Fig. 6.6:
Deflection is
reduced with a
double span as
shown in the
lower diagram,
and bending
strength is
increased ... but
shear strength
decreases by
about 25%.

A full explanation is lengthy (see pages 24 and 25 of *Timber Framing for the Rest of Us*), but it has to do with the severe tension forces in the wood fibers over the top of the center girder, as the two spans are loaded. These extreme tension stresses are actually eliminated by using two short rafters. At the Cave, we would have been better engineered had we cut the rafters in half and installed them as per the top of the diagram. It is true that we would have slightly decreased bending strength, but we were overbuilt that way and could give some back in favor of increased shear strength, where we were lacking. Deflection was not a problem, either, as our planking would be exposed above the rafters. With sheetrock ceilings, tape and spackle, deflection becomes more of a consideration, because cracking can develop in the spackle and paint.

And, it would have been easier to install two of the shorter rafters than the single heavy 18-footers. They were particularly heavy because our

rafter order had been delayed at the sawmill and the timbers were still in the form of logs when the girders were installed. And hemlock has an exceptionally high moisture content in its green state, compared with its seasoned state. The lesson here is to have materials lined up well in advance of need. Some small-time sawyers, who might approach the milling of lumber as a part-time job or a hobby, take a lackadaisical attitude towards things like delivery. Even a few weeks of drying makes a huge difference with regard to moisture content in lumber, and, therefore, weight and dimensional shrinking. Fortunately, lumber doesn't shrink significantly on length.

We made one final error that further diminished shear strength of the rafters: we "birdsmouthed" the underside of the rafters where they would bear on the girders and walls, as seen in Figure 6.7. This is actually fairly standard building procedure: on a pitched roof, the birdsmouth transfers the load nicely from above onto the bearing wall or girder below. The problem is that notching into the wood by even ¾-inch, in the case of the Cave rafters, diminishes the effective cross-sectional area (and the shear strength) of the member at that critical point by about 10 percent. On shallow pitches of up to 1.5:12, it is easier and stronger to shim with wooden shingles of the same width as the rafter. It might take two to four tapered shingles to match the pitch of the roof. Also, eliminating birdsmouths takes Mr. Murphy out of the picture, at least for the moment. It is all too easy to put the birdsmouth in the wrong place on the timber, thus spoiling a fairly valuable stick.

In 1977, I had no understanding or even awareness about shear stresses. My approach was to overbuilt on bending and figure that would do the trick. So, thanks to ignorance, I made three errors in our rafter design that affected shear strength: (1) I used hemlock, notoriously low in shear strength; (2) I used a supported double-span configuration, whereas two shorter rafters would have been stronger on shear; and (3) I birdsmouthed the rafters, whereas wooden shingles for stability would have been stronger.

Years later, when I had learned how to do stress load calculations for shear and bending, I did the numbers on Log End Cave, as built. On bending, the rafters and girders were way overbuilt, as I expected. On shear, the girders were good, but the rafters were, theoretically, underbuilt by about 10 percent. It would have been easy to attend to that shortfall by avoiding any of the three errors listed above. On the positive side, when you engineer to the numbers, there is a safety factor of six or seven built in. This is proper engineering, and should more than take care of a 10-percent discrepancy. Nevertheless, always work to the numbers and have the plans checked by a qualified structural engineer.

The 40-by-40-foot plan, given in Chapter 1, corrects these deficiencies.

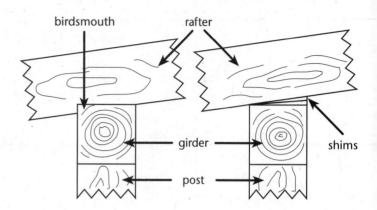

Log End Cave: Installing the Planking

The Cave plan had twelve sets of parallel rafters, creating eleven planking spans, as per Figure 6.10. Our choice of 32-inch centers for rafter spacing is a bit unusual, and we had to allow for an extra foot for any planks that ran on by the north and south rafters for overhang. If you are cutting your own trees for delivery to the sawmill, or if someone is supplying logs for you, it is good to cut the logs into useful lengths. With 32-inch centers, there is a lot of potential for waste. I gave my logger a list of useful lengths, as per Figure 6.10, but make sure that the logs are at least three inches longer, so that you can square the planks and eliminate bad ends. "Store bought" lumber comes in finished lengths (8-foot, 10-foot, 12-foot, etc) and have nice squared ends. These lengths, too, can be planned to minimize wastage. An eight-footer is a perfect 3-span, for example, while a 12-footer will do as a "Long 4," with just four

Fig. 6.7:

Left: Birds-mouthing decreases cross-sectional area of a beam, and, therefore, its shear strength. Right: Shimming does not decrease the shear strength of a beam.

Installing the Rafters at Log End Cave

After snapping the photo Figure 6-8 , I moved to the center girder (not shown) so that Bruce and Steve could slide the rafter up to me.

We waited until all the rafters were in place – half are shown in Figure 6.9 – before fastening them with toenails. Since the mid-1980s, I have used screws instead of nails for this purpose. And, now, in the new century, I use 12-inch-long timber-framing screws made for the purpose, such as TimberLok™ and structural screws from GRK Fasteners, both listed in Appendix B.

With rough-cut rafters from a local mill, there is likely to be a variance of a quarter-inch or more on the depth of the timber. As we want all the tops of the rafters to be in the same plane for a good flat deck of planking, we level the tops by installing additional wooden shingles, as necessary, where the rafter rests on girders or the east and west side walls.

Fig. 6.8: We slide a rafter along the girders.

Fig. 6.9: Log End Cave, with half of the rafters in place.

inches of waste. Your project will be different, so plan accordingly.

Start at the bottom and work up. Get the first course of planking absolutely straight – use a chalk line – and check that each end of the first course is the same distance from the rafter peak. We actually had some 18-footers, cut from the same logs that our four-by-eight rafters were

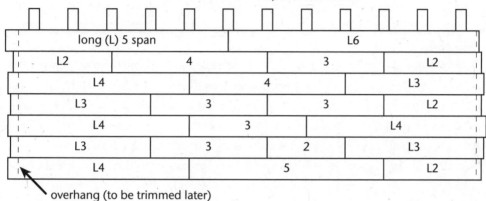

SPANS	LENGTH	SPANS	LENGTH
2	5' 4"*	4	10' 8"
L 2	6' 4"	L 4	11' 8"
3	8' 0"	5	13' 4"
L 3	9' 0"	L5	14' 4"

* too short for the sawmill; cut fours in half

overhang (to be trimmed later)

Fig. 6.10:
Plan ahead: cut logs and planks to useful lengths.

made from, and we started the first course on each side with these, so that a straight course could be established with just two planks.

Our wood was as "green" as could be, so we nailed the planks as shown in Figure 6.11, below. This method allows the nails to bend more easily, so that the wood can shrink without splitting. Resin-coated 16-penny (16d) nails work well, because they hold fast into the rafters. And their narrower shank makes splitting less likely near the end of a plank, and helps the nail to bend, if required, because of wood shrinkage.

We knew these boards were going to shrink, so we wanted to make sure we had them as tight as possible to begin with. Also, boards are not always perfectly straight, and you may have to

pull them in tight together in some way. An old carpenter's trick shown in Figures 6.12 and 6.13 works well. Use a 10- or 12-inch log cabin spike – or a long "star" chisel – as a lever to pull the boards tight against each other. Start the nails as shown, then whack the point of the spike about ⅜-inch into the rafter, tight against the board. Then, using one hand to pull the spike towards you, like a lever, hold it there while you drive the

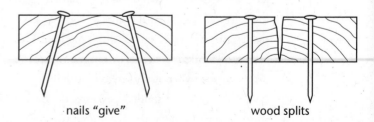

nails "give"   wood splits

Fig. 6.11: A carpenter's trick: angling nails allows for wood shrinkage without splitting the board.

Delays are a Part of Building

Owner-builders are often about 25 percent short in their cost estimates, but, when it comes to the time that things take, they are more typically 100 percent off, or worse. Everything takes longer than you can possibly imagine. Sawmills are late on orders. Mr. Murphy sneaks up with mischievous surprises. Someone (like me) gets hurt and loses a week of production.

The rafters were completed on August 26, completing a 53-day productive period that had begun on July 5, when the front-end loader took the first big bite out of the knoll. Then I fell off a rafter onto the 4-by-10 window header below, badly bruising, if not cracking, a rib. Fifteen minutes later, I was back up there, but being a whole lot more careful. But five days later, I couldn't do any physical work and the project came to a halt.

Another problem was that we were about 15 percent short on the 2-by-6 roof planking, the fault of our log supplier, not the sawyer. As it was September, I couldn't afford to wait for the unreliable supplier and purchased six plantation pines from a neighbor and had them hauled to the sawmill for immediate cutting and planing. Later, we mixed them with the hemlock planking. This worked out okay, but, again, they were not seasoned, so there was more shrinking than I would have liked. Fortunately, with wood heat below, they dried over the winter. At Earthwood, I forked out the extra bucks for kiln-dried two-by-six tongue-in-groove material.

I mention all this to let you know that building one's own house is not always clear sailing. And we were dealing with methods about which there was very little information at the time.

nails in with the other. Drive the nails angled towards you first, then the others. You can even tighten two courses at a time by this method. This method works well on tongue-in-groove planking, too, when it is giving you trouble.

At the north and south edges, leave the boards sticking out an inch or two more than your planned overhang. Later, you can snap a chalk line and cut the edge off all at once with a circular saw. Because of the green planks we used at the Cave, I actually trimmed the overhang with a chainsaw, as 1¾-inch-thick wet wood will choke all but the toughest circular saws.

 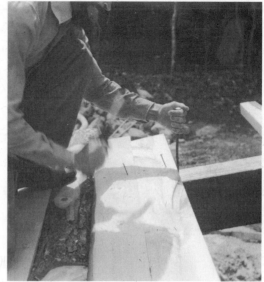

Fig. 6.12 & 6.13: For a tight fit, nail planks as shown in this sequence.

Pictures of Log End Cave show that we used two-by-twelve fascia boards to retain the earth on the north and south gable ends, something I wouldn't do again, as it is a very tricky detail to build, to waterproof, and to flash. Instead, I recommend the use of either pressure-treated retaining timbers, or, better, moss or grass sods to retain earth. Both methods are described in Chapter 8.

At the shallow "peak" of the roof, it is unlikely that full-width boards are going to miraculously fill the last spaces perfectly. You will almost certainly have to rip the edge of a board to make it fit, probably on both sides. You might even put a bit of an angle on your rip, so that the boards make a tight and tidy fit at the top.

EARTHWOOD: TIMBER FRAMING

The easiest and best way to lay out floor joists and rafters in a round house is with a simple radial rafter system. I have seen people try to marry other roof systems to the round design, like hip roofs (which are great for square buildings), but I have yet to see a successful example, so I am not going to waste the reader's time with these schemes. Radial rafter systems also work well with 16-sided buildings and octagons, as we have found at Stoneview, our new guesthouse. I would think that they would also work well with 6- and 12-sided buildings, too – the principle is the same – but I have no personal experience with those.

With a small round building, like our 16-foot-diameter library or 20-foot-diameter office building, both seen in the color section, a single

*Fig. 6.14:
This capital,
made for the
new earth-
roofed
Stoneview
guesthouse at
Earthwood,
allows 16 rafters
to bare
comfortably on
a round cedar
post.*

post can act as the high-end support at the center of the building. A small post, of course, even one a foot in diameter, doesn't have enough bearing surface to receive sixteen heavy timbers. On several buildings I have drawn from Greek architecture and broadened the top of the post by the use of a capital, an example of which is seen in Figure 6.14.

But, with large earth-roofed buildings, like Earthwood itself, the excessive span from center to perimeter dictates the inclusion of an inter-mediate support structure. Remember, cutting span in half makes the structure – not *twice* – but *four* times as strong.

An octagon works very nicely as this intermediate support structure. But, as already stated, my lifelong obsession with billiards necessitated the elimination of a post, accomplished with a very large (10-inch-by-12-

inch) oak girder. So we have, if you like, a truncated seven-sided figure as an intermediate framework, seen back in Figures 1.16 and 1.17. I suppose it is strictly a heptagon, but, as a former geometry teacher, I prefer to think of it as a "truncated octagon."

As stated, posts are the strong component of a post-and-beam frame. For five of our seven posts, downstairs, I used sound 8-by-8 barn timbers. For the two posts supporting the heavy oak girder, I had new 8-by-8 oak posts milled to match.

As at the Cave, we put a damp-proof course between the post and the concrete slab, but did not "pin" the posts to the floor. Six of the seven girders downstairs were also old barn timbers of similar dimensions. The other, of course, was the oak behemoth. The best of these recycled girders were used where unsupported clear spans were called for; those with mortise notches were used where internal framed walls would rise up from beneath and lend support.

The girders are fastened to the tops of posts with toenails, "toe screws," or right-angled metal fasteners made for the purpose. The tops of the girders are fastened to each other with truss plates, as seen in Figure 6.15. We planned the post lengths and girders so that the top of the girder ring would be 7-foot 3½-inches, corresponding to an exterior wall of 11 courses of 7⅝-inch blocks and a single course of 3⅝-inch solid blocks. Later, the load-bearing stone mass at the center of the building was also built with a shelf at the 7-foot 3½-inch level. With all supports at the same height, the floor joists would be level and in the same flat plane.

The Masonry Stove Central Stone Pillar

All support components of the home must proceed together, so that the floor joists and second story can be built. At Earthwood's center is a 23-ton cylindrical stone support pillar, which doubles as a masonry stove (or Russian fireplace). The diameter of this large stone pillar is five feet through the first story, reducing to four feet through the second story. This reduction in diameter creates a six-inch shelf for the floor joists to bear upon. . I designed the stove to meet our structural requirements and it was built by Steven Engelhart, a friend who had built several other masonry stoves before ours. We had lots of free indigenous stone and sand on site, and needed only to purchase fire bricks, flue tile, and cement. We labored for Steven, mixing mortar and supplying him with stone on the scaffold. With what I now think of as a Herculean effort, Steven built a beautiful stove for us in 54 hours. I know that because I paid him by the hour.

The building of the stove is outside of the purview of this book on earth-sheltered housing, but those interested in a detailed captioned visual presentation of its construction can look under "Masonry Stoves" in Appendix B..

EARTHWOOD: THE FLOOR JOISTS

Despite our experience at the Cave, the floor joists at Earthwood are four-by-eight hemlock.

truss plate

Fig. 6.15:
At Earthwood,
truss plates tie
the girders
together over
the posts of the
truncated
octagonal
frame. This
internal frame
cuts joist and
rafter spans in
half.

Floor loads are very much less than heavy earth roof loads, so shear doesn't come into play. And I already had the timbers ... and they were well-seasoned.

There are sixteen primary joists – each of them at least 18 feet long – which are supported by the perimeter walls (surface-bonded blocks or cordwood) at their outer ends, Steven's stone mass at the center, and the truncated octagon framework at mid-span. Actually, 11 of the 16 primary joists are 21 feet long, with the extra three feet cantilevered over the outside wall to support a walk-around deck about two-thirds of the way around the building. There are also sixteen secondary rafters which span from the exterior walls to just beyond the octagon interior post-and-beam framework. Planking spans are very short within the octagon, so every second rafter can be eliminated within its confines. Again, a glance at Figure 1.17 will clarify this point.

Wooden Plates on the Cordwood Wall

Like the masonry stove, cordwood masonry is beyond the scope of the present volume. Readers interested in cordwood are referred to my companion volume, *Cordwood Building: The State of the Art* (Appendix B).

Below grade, the joists rest on the course of 4-inch solid blocks with only a damp-proof course of plastic between timber and block. On the southern hemisphere, floor joists are supported by

Fig. 6.16: Jaki fastens the wooden plates to the four-by-eight window lintels with 16-penny (3¼-inch) nails. To stop draft, she has installed a strip of fiberglass insulation between the plates and the lintels.

the cordwood walls, which are capped with a series of two-by-six wooden plates, seen in Figure 6.16. As this system of plates is particular to the Earthwood design, I will explain them here.

As with all of the joist support components, the 2-by-6 wooden plates are placed on the cordwood wall so that their top surfaces are all at 7-foot 3½-inches off the slab. On the first story, there are 16 pairs of plates, not 32, because the earth-sheltered part of the northern hemisphere is continuous concrete blocks. But, later, when it came time to do the second story plates, we installed a full complement of 32 pairs of plates, one for each of the primary and secondary 5-by-10 rafters.

The plates are designed to distribute the concentrated load of the joists (and, later, rafters) into the cordwood wall. Such a point load, undistributed, could cause severe stresses on the cordwood masonry walls. Think of a knife cutting into a block of cheese.

Exterior plates are 3-foot 7-inches long and interior plates measure 3-foot 5-inches. The different lengths accommodate the differing circumferences, inner and outer. In most places, we were able to nail the plates right into double-wide 4-by-8 window and door headers or lintels, making them, effectively, 4 inches deep and 16 inches wide. During cordwood construction, window frame heights have to be planned with these lintels in mind. We used thin strips of fiberglass as a sealer between the plates and the lintels, but you can also use ¼-inch foam Sill

CHRIS DANCEY

Left: Chris and Wil Dancey's home in Ontario.

Below right: Chris Dancey's living roof.

Below left: The author's original Log End Cave, in summer and winter.

ROB ROY

CHRIS DANCEY

ROB ROY

The earth-sheltered home of former Earthwood students Geoff Huggins and Louisa Poulin was built in the '80s near Winchester, Virginia, using the techniques described in this book. Cordwood masonry is featured on the south side and half of the east side.

ROB ROY

ROB ROY

ROB ROY

Above: From the north and east, Geoff and Louisa's home is hard to spot, appearing as a grassy hillock.

Left: Stained glass flower designs, set in thermalpane, make beautiful snow-blocking between rafters. Log-ends are Virginia pine.

Top and below: The library and office buildings at Earthwood Building School in West Chazy, NY have living roofs of 6 to 7" topsoil. Moss and grass sods retain the earth at the edges.

ROB ROY

ROB ROY

Right: The office roof was lush with wildflowers for a couple of years, 26 varieties according to the label on the seed package.

Top far right: The post and capital at the center of the Stoneview guesthouse at Earthwood supports a considerable portion of the living roof above.

Bottom far right: The roofing detail at Stoneview guesthouse is described in Chapter 8.

ROB ROY

Above and inset:
Mark Powers and
Mary Hotchkiss'
40 x 40' Log End
Cave-style home in
Michigan, completed
in August of 2005.
(See Chapter 12.)

Right:
Mark and Mary
in their new home.
The engineered
paralam beam
overhead spans
18 feet.

Rob Roy

Jim Milstein Inset: Rob Roy

Above and inset: October in Pagosa Springs, Colorado is luxurious and comfortable in "The Asterisk," designed and built from 2000-04 by Jim Milstein, using the shortcrete method developed by Dale Pearcey at Formworks (Appendix B). While this home is outside of the cost parameters of this book, it is inspirational and shows that expensive homes too can be beautiful, environmentally harmonious and energy-efficient.

Left: The interior of Jim Milstein's home. Jim calls his house "The Asterisk" because three barrel-vaults intersect to resemble the punctuation mark.

Right: Plants are grown directly in Enkaroof VM and transported to the site, ready to unroll.

Center left and right: Sedum choices are almost endless in green roof applications.

COLBOND INC.

COLBOND INC.

COLBOND INC.

Bottom left: The Enkaroof VM system from Colbond.

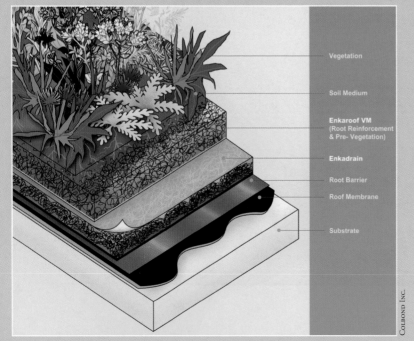

Vegetation

Soil Medium

Enkaroof VM (Root Reinforcement & Pre- Vegetation)

Enkadrain

Root Barrier

Roof Membrane

Substrate

COLBOND INC.

Seal™ at this detail, nicer to work with. In the very few instances where there was no header to nail into, we would set several roofing nails on the underside of the plates, leaving them sticking out about ⅝-inch. These plates would go in at the same time as the last course of cordwood mortar. The nails grab the mortar – or vice versa – very much reducing the likelihood of plates being dislodged when the joists or rafters are installed.

Again, rough-cut floor joists and – later – roof rafters are not likely to be all the same height. And there could be a little variance in plate height, too. To get the top surface of joists in the same flat plane, judicious use of handy wooden roofing shingles works very well.

EARTHWOOD: DECKING

Both the first floor and the ceiling/roof of the second floor are made from two-by-six tongue-in-groove kiln-dried spruce. There is a V-joint on the side of the planks which are exposed to the ceiling, an attractive detail. The topside of the first floor, and the roof deck, feature the boards tight against each other. You don't want the V-joint exposed on the top surface, which would make the floor hard to clean, and, on the roof, make the deck less amenable to application of the waterproofing membrane, the subject of the next chapter.

Laying planking on a 16-faceted radial rafter system is not really much more difficult than installing it over parallel rafters, although it takes more time. There is not very much waste, because

good use is made of short scraps near the center, as seen in Figure 6.17. There are three different situations that arise, which are:

The first facet. Start the first triangular facet at the outside of the building with full-length boards of an appropriate length. (We found that 12-footers on the first floor worked very well, but that 16-footers were more useful on the roof, because of the extra wall thickness that had to be covered, plus the 24-inch overhang.) Let one end of the board be just a wee bit longer than needed to cover a chalkline snapped down the middle of the joists. Let the other end of the board run on by the last beam in that facet. After you've got about half way to the center, or maybe a board more, snap

Fig. 6.17: Jonathan Cross nails the two-by-six planking to the radial floor joist system. Even short scraps are useful.

chalklines on the planking, directly over the centerline of the outermost beams in that facet. Then, set your circular saw about ¹⁄₁₆-inch deeper than the thickness of the planking and cut the edges. You'll have a nice straight cut, the saw blade barely tickling the timber below. Now, take the scraps you have cut and use them in inverse order to complete the facet right to the center. Again, snap lines, and trim the edges with a single cut.

Facets 2 through 15. (Or, if an octagon, facets 2 through 7.) Again, we start at the outside, with the longest boards. One end of the plank needs to be fitted carefully to the angle formed by the first facet. Use an adjustable angle square, sometimes called a bevel square, to find this angle. Then transpose the angle to the piece you're cutting, mark it with a pencil, and cut it with your circular saw. Nail the plank down, leaving the other end to run on by the last member of that facet. As with the first facet, use full length boards until you are about half-way to the center. Then snap a line and trim the facet edge, all at once. Again, use the scraps in reverse order – longest first – and continue to the center. Snap and cut. There won't be much waste.

The last facet. (This would be Number 16 at Earthwood, Number 8 if you are building an octagon.) Because of the tongue-in-groove feature, you cannot start at the outside and work in, as you have on all the other facets. Topologically, the board will not fit into the trapezoidal space defined by the two adjoining facets. This is where you need to show a little faith, tempered, of course, by all the experience you have gained on the other facets.

From a full-length board, cut the first small trapezoid you need, the piece closest to the center. I know. It hurts to do that. This is the faith part. Set the long scrap aside, in an ordered way. You will use it when its turn comes up further along the facet. All the pieces that you are cutting and fitting in this last facet are trapezoidal in shape. Measure the long end, which is the space between corresponding adjacent facets. With your adjustable angle square, transpose the angle from each end of the space over to the board you are cutting. Do not assume that the angle is the same at each end. That would be nice, but by now you have met the mischievous Mr. Murphy. The fresh-cut board fits in nicely from the exterior (open) edge of the facet. Keep doing this, and, finally, when you can get a useful piece from the smallest plank in your scrap pile, start using the ordered scraps. The shortest remaining piece should always work, until, finally, the longest scrap will finish the facet. Have an extra few boards available to take care of faulty pieces.

EARTHWOOD: UPSTAIRS POST AND BEAM FRAME

There is an almost identical post-and-beam frame upstairs. It is critical that all of the seven primary upstairs posts are superimposed directly over their downstairs counterparts. The roof load (also called "the line of thrust") is transferred down by compression through intermediary girders and joists and directly onto the post below. The worst possible design would be something that looks like the right side of Figure 6.2, just begging shear failure.

Where the framework differs is over the super-girder down below, the ten-by-twelve clear oak. Upstairs, we used an attractive 15-foot eight-by-eight barn timber in the corresponding location. It probably has between a third and one-half of the strength of the big guy downstairs, but that's okay, because we used three additional barn timber posts along its length, and interior walls between them, so the smaller girder's load is distributed equally along the oak girder below. And the oak timber is engineered to carry the heavy earth load.

EARTHWOOD: RAFTERS

The rafters are mostly five-by-ten red or white pine, or equivalent. I prefer the white, as the red pine trees tend to grow spirally, which can cause twisting of the rafter in place. When specifications say "or equivalent," this means "of equal strength." Because the octagon post-and-beam frame is truncated to a heptagon, about a quarter of the rafters upstairs have slightly greater spans. For this reason, six-by-ten rafters are specified in these locations.

The roof pitch at Earthwood is about "one in twelve" (expressed as 1:12). Pitch measures the "rise" of the roof (the first number), compared with its "run" (the second number, usually 12), the horizontal distance between one end of the pitch

Fig. 6.18: Planking the roof at the Stoneview guesthouse was a lot like doing the Earthwood roof. Here, we snap a line on the planks, but directly over the middle of the 4 × 8" rafters. Next we cut off all the "waste" at once with a circular saw.

Fig. 6.19: The "waste" from the first cut is used further up the facet, so it isn't waste at all. Next, we would snap chalklines again and trim the two saw-toothed edges.

and the other. I like any pitch between 1:12 and 2:12 for earth roofs. Within this range, the roof has enough pitch to prevent standing water, but not so much that the earth wants to slide off.

Trim the overhanging rafters to length before beginning the planking. Planking is as described above under Earthwood: Decking. Let the first board, all around, extend at least an inch beyond the trimmed rafters, so that the drip edge you'll be installing later (Chapters 7 and 8) keeps the dripping water off of the rafter ends. Figures 6.18 and 6.19, above, show the roof deck at Stoneview being installed.

Chapter

7

Waterproofing, Insulation & Drainage

The chapter title can be thought of as the mantra for installation. The correct order of events is waterproofing first, then insulation, then drainage. This is true for sidewalls (discussed first) and on the roof. On the east and west sides of a Log End Cave-type of building, the wall protection system melds in with the roof protection system as a continuous unit.

How we waterproofed at Log End Cave (lots of black plastic roofing cement covered with 6-mil black polythene) and what I now recommend are two entirely different things. What we did worked, but I don't advise it and wouldn't wish the job on anyone I want to keep as a friend. Essentially, and unknowingly, we were trying to invent Bituthene®, described below, and doing a sloppy job of it.

Waterproofing Options

Since building Log End Cave, we have used W.R. Grace Bituthene® waterproofing membrane, and similar products, for waterproofing walls ... and roofs. (Since the Grace patent ran out, several other

companies have been producing the same sort of membrane.) There are other waterproofing options, discussed briefly below, but for the cost parameters that I set (which is to own the home myself instead of the bank owning it for me), and for ease of application by inexperienced owner-builders, these self-bedding sheet membranes are a joy.

Here are some other "waterproofing" options you may have heard of:

Black tar coatings. These are damp-proofings, not waterproofings. I have not heard of anyone in the earth-sheltered housing field who considers such a coating to be a long-term solution.

Thoroseal™ (Thoro® Corporation) and other cementitious coatings. This class of foundation coating, made by mixing cementitious powder and water, is often specified by lending institutions to be applied to the sides of basement foundations. It works like a ceramic glazing, and is very waterproof indeed ... until it cracks. We used it at the basement of Log End Cottage, over mortared up blocks. Once the third course of blocks shifted

– due to improper drainage and insulating techniques – the Thoroseal™ layer cracked and let the water in. Again, the product has its uses, but should not be considered a waterproofing membrane for earth-sheltered housing. Similarly, surface-bonding cement, with plenty of calcium stearate in it, is also an excellent waterproofing, until it cracks. I consider it to be an excellent base for the true waterproofing membrane to follow, giving extra protection.

Liquid-applied membranes. There are lots of manufacturers, and some are listed in Appendix B. I have no personal experience with these membranes, although I have smeared hundreds of gallons of black plastic roofing cement all over Log End Cave. The difference is that these liquid-applied membranes do become dry and flexible. The tricky part is to apply a consistent and adequate thickness to the wall. These can be useful if you have an oddly shaped substrate, such as a truly spherical dome, where sheet membranes won't lay down nicely to the surface, or for certain detailing.

Bentonite clay membranes. Bentonite is a naturally occurring expansive clay. When water molecules try to work their way through the layers of clay platelets, a great hydrostatic pressure is created, pressing the plates so hard against each other that the water can only penetrate a tiny bit before the clay seals off against further penetration. Several companies make waterproofing products based on refined bentonite

clay. For example, Infrabond (Carlisle Coatings and Waterproofings) is a quarter-inch-thick mesh infused with bentonite. You can also buy pre-mixed trowel-grade consistency products, usually used for detailing in combination with the sheets. Price is moderate. Application is best left to installers with experience in the system. On the plus side, it is the nearest thing we have to a natural waterproofing, although some applications require a polyethylene sheet as a protective coating. And it is the only membrane with the ability to reseal itself. A slight drawback is the weight of the rolls. The "small" sheet of Infrabond is 4½ feet-by-15½ feet and weighs 79 pounds. Natural builders should research further, as, again, I cannot contribute with personal experience. A good place to start is to Google search *bentonite + waterproofing,* which returns several useful websites. Manufacturers are listed in Appendix B.

Fully-bedded sheet membranes. Butyl rubber, EPDM, neoprene, and other thick flexible-sheet waterproofing membranes are available. They are excellent waterproofing membranes, but must be fully or partially bedded to the substrate with mastic. The rolls are extremely heavy and should be applied, whenever possible, in large monolithic sheets, so physically getting them onto the roof could be a problem, as is hanging them on sidewalls. These membranes use a lot of petrochemical products, are expensive, and are best applied by professionals. Some of the rubber membranes have excellent UV resistance, something to consider where exposure to sun cannot be avoided.

"Modified" roofing. There are some thick asphalt-based roofings which have been effectively used as waterproofing with earth-sheltered housing. The sheets are typically three feet wide and an eighth-inch thick, and one sheet overlaps the next by 4 to 6 inches. The weld between sheets can be torched down, or chemically welded. Architect David Woods, who is also Secretary of the British Earth Sheltering Association, strongly recommends this type of membrane, but cautions that it is more expensive and more work than the self-adhering membranes, and advises that it be installed by experienced professional roofers.

Layers of roll roofing alternating with hot pitch. This old-fashioned method is definitely not recommended as a long-term waterproofing below grade.

In Search of the Perfect Membrane …

A perfect membrane would be "all natural," have low embodied energy, be cheap and easy to apply, and give at least 100 years of trouble-free service. The closest I can think of is the way the Neolithic people in Scotland's Orkney Islands successfully waterproofed the 5,000-year-old Maeshowe burial chamber. The structure is a large corbelled-arch dry-stone burial vault, covered with a thick layer of good clay. The structure stayed waterproof until the Vikings broke in through the roof about 1,000 years ago.

Paul Isaacson, my underground co-instructor back in the Mother Earth News Eco-Village days of the early 1980s, told me that he successfully waterproofed his home in Utah by placing a tandem-trailer load of unrefined bentonite clay against the walls. That home no longer exists – the property was bought out and redeveloped – but Maeshowe has had its roof repaired and was still dry last time I went inside.

And the perfect membrane? Read on.

Bituthene® (and Similar) Membranes

The nearest thing I have found to the perfect membrane is the Bituthene® waterproofing membrane, manufactured by Grace Construction Products, listed in Appendix B, along with similar products manufactured by others. Bituthene® is very reasonable in cost (although this is bound to increase with oil prices), it is effective and long-lasting, and its application is easy to learn. Its drawbacks are: It must be protected fairly soon from the sun's harmful ultraviolet rays; it contains lots of petrochemical products; and it has been known to be penetrated by certain chomping insects. In 24 years, we have twice had to repair tiny holes in the Earthwood roof caused by carpenter ants. They penetrated from below, incidentally, not from above.

Bituthene® is composed of two cross-laminated layers of black polyethylene bedded over a uniform 60-mil ($\frac{1}{16}$-inch) layer of rubberized

Chomping Insects

Roger Danley and Becky Gillette wrote a case study for my earlier book, *The Complete Book of Underground Houses*. They built their home in southern Mississippi. Roger says:

"We got a deal on a large sheet of EPDM, a synthetic-rubber membrane, guaranteed for 40 years in an exposed industrial setting. Our only problem has been fire ants, who'll eat a hole right through a 120-mil (⅛-inch) sheet. So we have to be vigilant about treating any mounds we see on the roof. I don't think it would have made any difference which membrane we'd have used, they would have eaten through it. We mow about twice a year, so we can see any mounds that appear. We have a beautiful spring-to-fall wildflower meadow on top."

Becky adds: "(The fire ants) caused leaks in our EPDM membrane, and it's a pain to fix leaks on a sod roof. A leak doesn't always drip exactly where the hole is, so leaks are hard to find. You can dig up a large area to find a small hole. With 60 inches of rain a year, and as much as eight inches in a day, even small holes can be big problems. You can kill all the ants, and new colonies will spring up overnight. I don't know the answer."

Frankly, Becky, neither do I. I have yet to hear of insect damage through a quality membrane like Bituthene® fully bedded to concrete, so it is possible that a two-inch layer of concrete poured over the roof planking would provide an insect-proof receiving surface for the waterproofing. But the extra weight (about 28 pounds) would have to be engineered for. At least in southern Mississippi, there is no large snow load to factor in.

Got fire ants? Possibly a conventional roof is in order.

bitumen material. The bituminous bedding is very sticky, but the 36-inch-by-65-foot roll comes protected with a non-stick backing paper, removed during application. The factory edge of the membrane has a black sticky caulking on it, which protects the edge from lifting, a condition called "fish-mouthing."

WATERPROOFING VERTICAL WALLS

With designs where earth berms come right up and onto the roof, as at the Cave, the sidewalls must be waterproofed first, vertically (like wallpaper), but extended six inches onto the roof surface as well. Later, the roof is waterproofed with a horizontal application, overlapping the extended sidewall membrane by three inches.

Prior to application, the wall needs to be primed with the compatible primer that Grace supplies as a part of their waterproofing system. With the Bituthene® 4000 membrane, they supply a plastic jug of "surface conditioner" with each roll, enough to prime an area large enough for the roll. They do not supply the conditioner with their Bituthene® 3000 membrane, or their Bituthene® Low Temperature membrane. With the 4000 membrane, just prime as much wall as you intend to do on the day, and let it fully dry before installing the membrane. The conditioner goes on quickly with a roller and usually dries to a tacky feel within an hour. This conditioner should not be used with "green" concrete (less than seven days old). But, if time is critical, you can use a Bituthene® Primer B2 on slightly damp surfaces or green concrete.

The use of primer with the 3000 series is a little confusing, but the step should not be left out, even though local suppliers often carry the membrane, but not the primer. With Bituthene® 3000, for use with temperatures of at least 40°F (5°C), order and apply Bituthene® Primer WP-3000, which is similar to the one they provide with the Bituthene® 4000. In a pinch, you can prime the surface with DAP acrylic bonding agent or Acryl-60 bonding agent, usually readily available. For temperatures between 25°F (-4°C) and 40°F (5°C), use Bituthene® Low Temperature Membrane and the WP-3000 primer, except use the B2 primer over green concrete.

Bituthene® 4000, like Bituthene® Low Temperature Membrane, can be installed in temperatures as low as 25°F (-4°C) with the Surface Conditioner supplied.

With other manufacturers of this type of self-adhering membrane, be sure to check their temperature and surface preparation guidelines.

The ideal time to install Bituthene® – on wall or roof – would be a calm gray day in the 50° to 75°F range (10° to 24°C) On a cooler, but sunny day, you can work as low as 40°F (5°C). The sun on the sheets will help make the rubberized bitumen very sticky. Below 40°F (5°C), but not less than 25°F (-4°C) you need to use Bituthene® 4000 or Bituthene® Low Temperature Membrane, which stays sticky at these lower temperatures. On a warm sunny day, if a flap of membrane, with its backing paper already removed, happens to fold over and touch another exposed bitumastic surface, they fuse together instantly and irrevocably, never to be separated, and you have lost a few dollars worth of material.

With both poured and block walls, there is the potential for a leak where the wall meets the concrete footing, even if you bring the surface bonding cement out onto the little 4-inch-wide shelf of the footing which extends out from the block wall. Prime both the wall and the 4-inch exposed part of the footing with Grace Surface Conditioner or the 3000-WP Primer. Then, cut a 7-inch-wide by 36-inch-long strip of Bituthene® from the end of a roll. Remove the backing paper

Fig. 7.1:
The important
waterproofing
detail where the
wall meets the
footing is sealed
with a strip of
Bituthene®.
Here, I am
caulking the
freshly cut edge.

and press the strip carefully into the area where the wall meets the footing, as seen in Figure 7.1. In the picture, I have already pressed the strip into place, with about 3½ inches sticking to the wall and 3½ inches sticking to the footing. I am caulking the cut edge of the strip with a bead of Grace Bituthene® Mastic, to prevent fish-mouthing.

The factory edge of the membrane has a bead of mastic already in place to keep this edge stuck down, but exposed cut edges need to be caulked. I use the Grace Mastic, as other black plastic caulks may or may not be compatible with the membrane. Examples of incompatible mastic would include tar, asphalt or pitch-based

materials, or any mastic containing polysulfide polymer. We do not need to caulk the top edge of this strip, as it will be covered by 3 inches of Bituthene®, as seen in Figure 7.2. Later, the bottom edge of the vertical sheets will be caulked with the Grace Mastic.

Bituthene® is applied vertically on sidewalls. Cut the sheet to the correct length, making sure that the new sheet laps 2½ inches (manufacturer's recommendation) onto the previous sheet. Use the white guideline on the previous sheet to gauge this overlap. One person is positioned at the top of the wall, as per Figure 7.2, and another – preferably two others – are on the ground to receive the sheet and press it onto the conditioned surface. Try it with the backing paper still on it, and, when it fits right, with a 2½-inch lap over the previous sheet, mark the top surface on the wall with a pencil. Then, with the person below holding the sheet slightly away from the wall, the others pull the backing paper away from the top foot of the sheet. The ground crew keeps the edge of the new sheet lined up with the white line 2½ inches in from the edge on the previous sheet, and the person on top presses the foot of exposed bitumastic onto the wall, watching the pencil mark.

Now, one of the ground crew members can carefully peel the backing paper away from the underside of the sheet while the other presses it onto the wall. To eliminate bubbles forming behind the sheet, press first in the middle of the sheet and then work it right and left. By this method, the sheet should stay parallel to the previous one, maintaining the required overlap. Press hard with the heel of your hand to get a good joint at the overlap. Grace recommends a metal hand roller, but good pressure with the hand has worked well for us. Where the sheet overlaps the 7-inch filleted pieces already installed at the bottom of the wall, Grace says to "Press the membrane firmly to the wall with the butt end of a hardwood tool, such as a hammer handle. Failure to use heavy pressure at terminations can result in a poor seal." (Grace Product Data). Finally, caulk all cut ends with Bituthene® Mastic.

If it becomes clear that the edge of the new sheet is wandering off the line of the previous sheet – particularly if the overlap is getting smaller – you need to stop pressing the Bituthene®. You

Fig. 7.2: Bituthene® waterproofing membrane is applied vertically to walls.

Fig. 7.3: The Bituthene® has been applied to the lower story at Earthwood.

cannot stretch it back on course. Cut the sheet short and begin again with at least a 3-inch overlap over the cut-off edge. Always press terminations extra-hard for a good seal, and caulk any new cuts with Bituthene® Mastic.

The primary advantages of this type of membrane are speed, reasonable cost of application, and quality control. In May of 2005, Bituthene® 4000 with conditioner cost about 85 cents per square foot, and Bituthene® 3000 without primer was about 70 cents. Any shape of piece can be cut and used to patch or fill an area, but always allow a 2½-inch overlap over previous sheets, overlap 3 inches over previous terminations, and seal all cut edges with compatible mastic. Figure 7.3 shows the lower story of Earthwood waterproofed.

WATERPROOFING ROOFS: THE DRIP EDGE

A "drip edge," on all exposed edges of planking, must be installed before the waterproofing. It serves two important purposes. First, as its name implies, the metal drip edge, extending a half-inch below the planking, causes water to drip off of its bottom edge, stopping it from running back along the planking towards the home, where it could deteriorate planking, rafters, and walls. Second, by exposing about five inches of the metal drip edge on the roof surface, the Bituthene® membrane can be kept well away from the edge of the roof, thus protecting it from ultraviolet deterioration.

For earth roofs, I find it preferable – and cheaper – to make my own drip edge from aluminum flashing, as opposed to purchasing manufactured drip edge. Commercial designated drip edge comes in 10-foot sections, and is usually made from aluminum or galvanized iron. Some have a baked-on enameled surface. Most of it seems to be made for use with plywood, under the bottom course of a pitched shingled roof, and so is neither wide enough nor deep enough for earth roofs decked with two-by planking. I have had good success on a number of projects by making my own drip edge from 14-inch-wide aluminum flashing – a common item – cut down the middle to make two strips of 7-inch flashing. Roll the flashing out onto a flat surface, such as the roof, and make a pencil line (or chalkline) down the center. Score the line with a sharp razor-blade knife, using a long straight edge to guide the knife. (My 4-foot by 2-inch ruled straight-edge is one of the most-used tools in my kit, and is particularly handy for cutting Bituthene®, so I strongly advise owning one.) You will find that by flexing the 14-inch strip back and forth along the score line, it soon separates nicely into two 7-inch pieces. Be careful. The freshly ripped edge can be sharp, with nasty little slivers of aluminum to cut your hands.

To make drip edge from the 7-inch strip, work along the edge of a sharp-edged table – or the roof edge itself, if you are careful! Extend the *factory edge* of the metal two inches over the table or roof edge. A helper is useful to hold the strip in place. Then, with your left hand, start to bend the sheet over the edge. At the same time, with your right hand, use your thumb and forefinger to press the bend into a sharp right angle. Practice with a small piece. The process is not difficult, although you must really press quite strongly with the thumb and forefinger to get that nice right-angled edge.

Now, with roofing nails, nail the drip edge down to the deck, so that the two-inch dimension hangs over the edge, with five inches on the roof. Keep the nails about two inches from the freshly cut edge of your homemade drip edge. Later, the first course of Bituthene® will cover that ripped edge, and the nails by a half inch, leaving 2½ inches of aluminum flashing exposed on the edge of the deck. Press really hard along that edge for a good seal. By this method, the membrane, subject to UV damage, can be kept well back from the edge of the roof, protected by the various layers to follow, described below.

Waterproofing a Cave-style Roof

The Log End Cave and Earthwood roof designs present two different waterproofing application scenarios. At the Cave, the earth roof meets with the east and west berms. At Earthwood, the roof is "freestanding." It also happens to be composed of sixteen triangular facets, like the blunt end of a diamond.

With a pitched rectilinear roof, such as the Cave, it works well to roll the sheet out to its full length, backing paper and all, and then slide the sheet over the top edge of the vertically applied sheets that you've left extending six inches or so

Fig. 7.4:
The author bends two inches of the 7" wide flashing over the edge of the 2 × 6" deck.

onto the roof. With a south-facing Cave design, the Bituthene® is running north-south. Give it a full 3-inch lap over the top edges of the wall membrane (which have been turned on to the roof substrate), and make sure it is covering the roofing nails on the flashing at its north and south ends. Again, about 2½ inches of the flashing will show.

Once the sheet is properly aligned – it could be 30 feet long or more – have one person stand on one end to keep it from moving. A second person carefully rolls the sheet up again. The person standing on the end can shuffle forward about three feet, being careful to maintain the good alignment, and stand heavy while the assistant lifts the short end of the sheet, pulls a couple of feet of backing paper away from the sticky bitumastic, and folds the paper up close to where the person is standing. Carefully press this two feet onto the primed wooden deck, making sure of sufficient lap on the drip edge.

Fig. 7.5: The roll is laid out on the correct alignment, with the backing paper still on. Jaki (right) and Diane (left) have cut two feet of the backing paper away and are guiding it down to the primed deck. The author presses the membrane down with firm hand pressure.

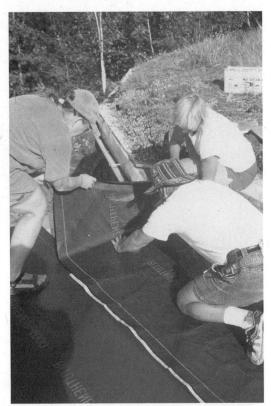

Roll the sheet back (with its backing paper) onto the part already welded to the deck, far enough that you can grab the folded backing paper from beneath the roll. Now, one person gets down on hands and knees on the fastened-down portion, while another grabs the flap of backing paper on the underside of the roll and begins to pull it away from the roll. Together, the two can unroll the sheet, one pressing the sticky surface down onto the deck while the other removes the backing paper and keeps the roll moving forward. The person pressing should smoothen the sheet from the middle to the outer edges, pressing firmly with the heel of his hands. Later, go over both edges again with the heel of your hand, trying to avoid the sticky factory-applied caulked edge.

If care has been taken to prevent the roll from shifting position during the steps described above, it will roll down on the original alignment, with the correct lap over the sheet (or sheets) down slope. If you can see that you are losing the correct lap, there is no option but to stop, cut the sheet off, and begin again with a three-inch lap over the portion already laid down. Weld this overlap with strong pressure.

Keep working up the slope with succeeding strips of membrane, always following the white lap line imprinted on the previous sheet. The roll, 66.7 feet (20m) long, will no doubt finish part way along a course. Simply start the new roll three inches onto the end of the previous roll. Sometimes, the last two or three feet of the roll is badly creased or wrinkled like the skin on the

Fig. 7.6: After the sheet has been rolled back up with its backing paper on, the team again unrolls it on line, pulling the paper away at the same time. The author, on hands and knees, presses the sticky surface of the membrane onto the wooden deck.

back of "dishwater hands." Better to discard these ends, because you might not get a good seal over the creases with the next sheet.

Do one side of the roof, as close as you can to the peak, and then the other side. If, at the peak, you are left with a non-waterproofed deck of 18 inches to 30 inches in width, you can lap both top edges of the uppermost sheets with a final full-width ridge piece. If the gap is quite small, say 3 to 12 inches, you could cut a sheet down the middle, yielding two 18-inch-wide rolls, and make your ridge piece from those. Bituthene® now comes with a "Rip Cord" feature that allows you to easily divide the sheet down the middle into two 18-inch pieces.

After all pieces have been pressed down to the deck, the drip edge, and to each other, caulk all cut edges with Bituthene® Mastic.

In May of 2005, in two hours and using the method described above, Jaki and I helped Bruce Kilgore install 800 square feet of Bituthene® 3000 membrane on a 1:12 pitch shed roof. He caulked all fresh cut edges a few days later.

EARTHWOOD: WATERPROOFING THE ROOF

Once again, work on a conditioned surface. Recondition any surface which has been rained upon, or has been exposed for 24 hours. Prime only as much deck as you can reasonably hope to cover in a day. Still, the work goes quite quickly, and you can get quite a bit done in a day. Because of all the individual trapezoidal pieces involved

with the large Earthwood roof, it is probably a bit ambitious for inexperienced owner-builders to expect to do the entire project in a day. A smaller straight-run roof like Log End Cave's is easy to do in a day.

The sixteen facets of the Earthwood roof is an entirely different situation from the Cave design, but, again, we begin at the bottom, with a lap onto the 5-inch-wide top surface of the drip edge. The first sheet will be just over 10 feet long. On the first couple of courses, you could use the technique already described above, but I prefer another method, even though it involves three people.

Roll the membrane out in place, and cut the trapezoidal piece you will need with the backing paper still on it. Cut over a scrap board, using your 4-foot steel straightedge to guide the razor knife, as seen in Figure 7.7, below. You may have a little gap between the decking on two adjacent facets, so make the piece long enough to extend three inches in each direction beyond the facet's saddle edge.

Fig. 7.7:
Cut Bituthene®
with a razor
knife, over a
scrap board.

Fig. 7.8:
Apply the lower
edge of the
Bituthene® first,
using the white
line on the
previous sheet
as a guide.

hand, shown in Figure 7.8. The carriers need to maneuver the low edge down to the drip edge, being careful not to actually touch it until it has just the right lap (½-inch clear) over the roofing nails. Then, with that edge straight, touch it down, the middle of the sheet first. The sheet will have a kind of a concave trough to it, along its length, which can be seen in the picture. This curve can be used to advantage, as you let the sheet "roll" itself out onto the surface.

Because you are working on the edge of the building on the first course, it is hard for the third person to help press the sheet down in the same way that he or she can on succeeding courses. For that reason, you may want to use the "roll it out, roll it back, roll it out again" method for the longest trapezoids. After the first course, though, the trapezoids are shorter and more manageable, and the third person can position him- or herself at the middle of the downside of the sheet and spread the sticky surface down, working it with the hands right and left, and helping to prevent bubbles or creases.

Later, when the adjacent sheets of the same course are installed, they too should extend onto the next facet by three inches, giving a six-inch lap where two facets meet. Normally, a 3-inch lap is sufficient with Bituthene® 3000 and 4000 membranes, but you want extra protection where facets come together, pressing hard onto three inches of solid wood on each side of the crack.

After cutting the trapezoid to size, turn it over and remove the backing paper. Now, carefully, with a person at each end, turn the sheet right side up and bring it over to where it belongs. The two carriers need to talk to each other about end lap, and a third person is useful to help guide the sheet into place, or to hold it in the middle as a helping

Work the first course all the way around the building, until the sixteenth sheet laps the first by six inches. The second course, away from the edge, is easier, because the third person can get down on hands and knees to ease the sheet out onto the substrate. Be sure to place the caulked factory edge right on the white guideline of the previous sheet. Again, the two carriers need to communicate well with each other with regard to lapping over the saddles of the facets.

The sheets get smaller and easier to handle towards the center. With experience gained, and the smaller sheets, the work goes faster, too. After all of the sheets have been installed, go around and caulk all of the cut edges with Bituthene® Mastic, as per Figure 7.9.

If you inadvertently create punctures or cuts, or see any suspicious marks on the membrane, it is easy to patch with pieces of the same material. Cut your patch so that it extends six inches in all directions from the damage, and press it down hard with the heel of your hand. A little trick I like is to cut the corners off a square patch so that it looks a little more like an octagon, lessening the chances of a corner lifting. Caulk around the patch.

SIDEWALL INSULATION

I use close-celled extruded polystyrene insulation below grade. It is readily available, and has a good insulation value of R-5 per inch. Use at least two inches (R-10) on earth-sheltered sidewalls in the North. At Earthwood, we used three inches (R-15) down to frost level, about four feet in our area. We did this partially because we had enough one-inch sheets left over to do it, but it is bound to keep the block wall's temperature more stable near the top of the berm. The insulation also protects the waterproofing membrane from physical damage and from freeze-thaw cycling, one of the main causes of roofing breakdown.

Expanded polystyrene, also called EPS or beadboard, may be cheaper than the extruded variety, but it is not closed-celled and is more prone to water penetration and, therefore, to loss of R-value. Also, some EPS insulations have a lower value to begin with, R-3 to R-4 per inch of thickness.

Polyurethane foam (R-8 per inch) is also subject to severe loss of insulation value if water penetrates any protective coating or surface, such as aluminum foil or sprayed-on protection coat.

With flat straight walls like Log End Cave, you can "spot glue" a four-by-eight sheet of 2-inch extruded polystyrene right to the wall. Generally six little globs of glue will hold the sheet in place. Make sure the glue is compatible with both the polystyrene and the membrane. Grace makes

Fig. 7.9: Exposed cut edges must be sealed with a compatible caulking. After applying a bead of caulk, use a butter knife to create an edge that looks something like the factory edge. Tip: bend the last inch of a non-serrated butter knife just a little bit to make an excellent tool for this purpose.

*Fig. 7.10:
Keeping an eye
out for large
stones, Jaki and
Dennis hold the
extruded
polystyrene
against the wall
during
backfilling.*

"Bituthene® Protection Board Adhesive" for applying lightweight protection boards, such as extruded polystyrene, to the membrane. Grace warns, however, that the adhesive is extremely flammable and advises installers to read their Material Safety Data Sheet (MSDS), available at <www.graceconstruction.com>, before using it.

With a curved building like Earthwood, you cannot glue these rigid sheets of insulation to the wall. The sheets will bend around the curve okay, but the glue won't hold them there. We had to physically hold the sheets up against the wall while we pushed good percolating sandy backfill up against them to hold them in place, as seen in Figure 7.10.

How much sidewall insulation to use was already covered in Chapter 1, under the headings "Insulation in the North" and "Insulation in the South."

ROOFING INSULATION

The quantity of roof insulation is a balance between you, your budget, local energy code and the open-mindedness of the building inspector. For the purposes of meeting energy code requirements, you may or may not get much R-value credit for having an earth roof. I maintain that the earth roof holds snow better than any other kind of roof, and that snow is excellent insulation. But such a statement will melt to water with a by-the-book code enforcement officer. New York State requires R-33 insulation on the roof, which would need six inches (R-30) of extruded polystyrene to satisfy. (The missing R-3 is more than made up by the 2-inch planking and anybody's reading of earth R-values.)

We have four inches of Dow Styrofoam® on the Earthwood roof, just R-20, but our home is one of the most energy-efficient homes in the North of New York. Yes, six inches might have made it even more efficient, but certainly not by 50 percent. You reach a point of diminishing returns. And the monolithic nature of the insulation – no conduction through rafters, as with other forms of insulation – helps a great deal, as does the minimal skin area of the building's fabric.

Your decision must be budget based, but, at the same time, you have to know that your plan will pass code, or, at least, your local building inspector. Let's be frank here: Building codes are a "one size fits all" type of thing. They don't easily accommodate sensible alternative methods of accomplishing energy efficiency goals, such as giving credit to the shape of the house, its orientation, thermal mass, or the fact that the home is is earth-sheltered or earth-roofed. The onus is up to the builder to make a good case for alternatives "which meet or exceed code requirements." And an engineer's stamp does carry a lot of weight, mainly because it takes the pressure off of the building inspector.

A Bituthene® type of waterproofing membrane needs to be covered as early as possible to protect it from damage from traffic and from the sun's ultraviolet rays. On the Cave style, just spread the rigid insulation over the single-pitched roof, using the largest sheets you can get, generally 4-by-8-foot. (Sometimes, only 2-by-8-foot sheets seem to be available. It is worth asking around.) Make use of the tongue-and-groove edge, available on some brands. Use the upper courses to fully lap all joints on lower courses. Choose a calm day, as it doesn't take a huge wind to blow rigid foam off the roof.

On many an earth roof, Jaki and I have placed our extruded polystyrene down on the membrane, and quickly placed planks, stones, blocks or other ballast to stop the lightweight sheets from blowing away. Ideally, we get the insulation itself covered as soon as possible with the drainage layer and earth, as described below, but there is sometimes a delay of a week or longer.

There are insulation/ballast products made by T. Clear Corporation that will facilitate the insulating process over a membrane. Lightguard, Coolguard, and Heavyguard are all 2-by-4-foot panels of tongue-and-grooved Dow Styrofoam® insulation, topped with latex-modified concrete bonded to the upper surface. Lightguard and Coolguard have a 3/8-inch coating of concrete, and weigh 4.5 lbs./sq. ft (36 pounds per panel), while Heavyguard has a 1-inch concrete top surface for a total weight of 11 lbs./sq. ft. (88 pounds per panel.) Coolguard, in addition, has a good reflective surface, as it is normally used without an earth cover. All of these products are available with Styrofoam® thicknesses of 2 inches, 3 inches, or 4 inches, and all can be applied over layers of regular uncoated extruded polystyrene. For living roofs, which will soon get drainage layers and earth layers, the Lightguard product is the most appropriate. In our part of New York, at least three schools have used Lightguard as the final roofing surface, but, for the earthern or living roof systems, you will want to make use of it as an easy-to-apply ballasted insulation. My suggestion is to find your best deal on extruded polystyrene – 4-by-8-foot sheets, if possible – and put down 2 to 4 inches of it before topping the whole insulation blanket with 2-inch panels of Lightguard. Tongue-and-grooved panels weighing 36 pounds each are not about to blow away. You could leave such a roof for months, even years, although it would be better to get the drainage layer on as soon as possible to take almost all business away from the membrane. While these

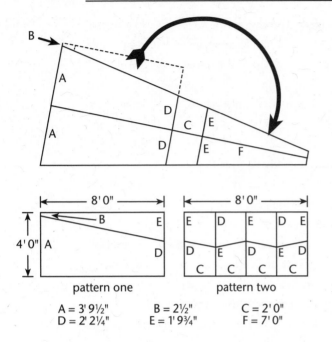

pattern one

pattern two

A = 3' 9½"   B = 2½"   C = 2' 0"
D = 2' 2¼"   E = 1' 9¾"   F = 7' 0"

Fig. 7.11:
The "no waste"
cutting pattern
used for roof
insulation at
Earthwood.

T. Clear products are very appropriate to large rectangular roofs with a single plane, or which meet at a peak, like Log End Cave, the multi-faceted Earthwood roof presents conditions that preclude their use.

EARTHWOOD: INSULATING THE ROOF

The installation of the Dow Styrofoam® on the 16-faceted Earthwood roof was greatly facilitated by working out two cutting patterns for the 4-by-8-foot by 2-inch-thick sheets that we had to work with. The patterns are reproduced in Figure 7.11. Each of the sixteen facets requires two sheets cut as per Pattern One, and a quarter of a sheet cut as per Pattern Two. Therefore, a sheet cut as per

Pattern Two will serve four facets, one layer thick. Thirty-six 4-by-8-foot sheets are needed to cover the roof surface with a single layer. As we used two 2-inch layers, we needed 72 full sheets to gain the 4-inch-thick (R-20) insulation layer we wanted. Add an extra layer for R-30. These patterns worked perfectly for our outside wall diameter of 38 feet 8 inches, and would be useful for house diameters of a foot larger or smaller. You would need to work out other patterns for 16-sided roofs, of significantly greater or lesser diameters.

The second course is offset two inches over the first, so that all the radial joints are covered by the second course. We actually made these special ship-lapped panels on the flat floor surface below and passed them onto the roof. We "tacked" them together with four to six 16-penny nails, and the nails stayed in place until all of the ship-lapped units were installed on the roof. These nails miss poking through the lower sheet by about ⅝-inch. Nevertheless, I didn't like the idea of hundreds of nail points aiming towards my membrane, so prior to covering the insulation, we went around on our hands and knees and pulled every single one of them out. Figure 7.12 shows the installation of these ship-lapped panels.

The overhang doesn't really need to be insulated. Nevertheless, we used an inch of Dow Styrofoam® over the membrane on the overhang as a protection board. And, this insulation, in combination with the 2-inch planking below the membrane, effectively sandwiches the membrane and diminishes the number of freeze-thaw cycles, and their severity.

The insulation should be covered as soon as possible with the all-important drainage layer and earth. If this is not possible, make sure there is enough ballast to keep the insulation from blowing away. How much ballast? Fully twice as much as you think you need. Maybe more.

DRAINAGE

Drainage is the better part of waterproofing. Water looks for the easy route. Give it an easier place to go than into your house, and it will take that easier route. I use the italics because I think of this as a mantra, and subject my students to it at least three times during our Underground Workshops.

After the insulation is in place, we can attend to drainage, beginning with the footing drains (also called French drains) at the base of the wall. Before commencing the drains, make sure that your wall insulation is continuous with the footing insulation, as per Figure 7.13. This will mean placing some insulation on the top of that little footing shelf that we covered with Bituthene® back in Figure 7.1. Any uninsulated part of the concrete foundation will be an energy nosebleed and cause condensation on the corresponding interior surface.

FOOTING DRAINS

The business of a footing drain (also called a French drain) is to carry water away from the important detail where the wall meets the footing.

Fig. 7.12: Jaki installs the shiplapped insulation panels on the Earthwood roof.

But, providing that the rest of the sidewall has good drainage as described below, it will also serve to carry away all of the water which would otherwise stand against the sidewalls. Footing drains consist of a flexible plastic perforated tube, usually 4 inches to 6 inches in diameter, surrounded by crushed stone. The tube slopes gently around the building, dropping no more than an inch in ten feet. At its high point, the top of the tube should extend an inch or so above the top of the footing. By the time the tube slopes to its low point, the top of the tube could be four to six inches below the top of the footing.

The crushed stone surround is kept clean from earth by covering it with a filtration layer. I have had good success with a natural filtration mat made by scattering 2 to 3 inches of loose hay or straw over the stone. This organic material decomposes into a thin but effective natural filtration mat, which keeps earth from entering the crushed stone and perforated drain tube. You can also use a manufactured drainage mat made for the purpose. These are usually made of nylon, polypropylene, or some similar petrochemical composite.

Figure 4.2 shows how the footing drains carried water away at the original Log End Cave.

The major change I would make is to take the drain out above grade, if possible, instead of to a soakaway pit, which can fill up and become useless. Figures 7.14 and 7.15 show the footing drains being installed.

DRAINING THE SIDEWALLS

Figure 7.13, with its key, describes the various components that make up the drainage system where a berm rises up and meets the roof. The drawing assumes that the backfilling material has moderate percolation characteristics. Note the use of drain tubing near the top of the wall, to gather water collected by the roof drainage layer, and another horizontal drain halfway up the wall. These secondary drains are connected to the footing drain by the use of vertical tubing, which can be non-perforated.

Sidewall drainage strategy depends on the percolation qualities of the backfilling material. Coarse sand (or mixed sand and gravel) makes an excellent backfill, allowing water to percolate down to the footing drain, where it gets carried away to some point above grade. Sandy backfill can be tamped in layers up against the wall, using a hand tamper or – gently – with the bucket of a backhoe or front-end loader. Do not use a powered vibrating compactor for this job: not only

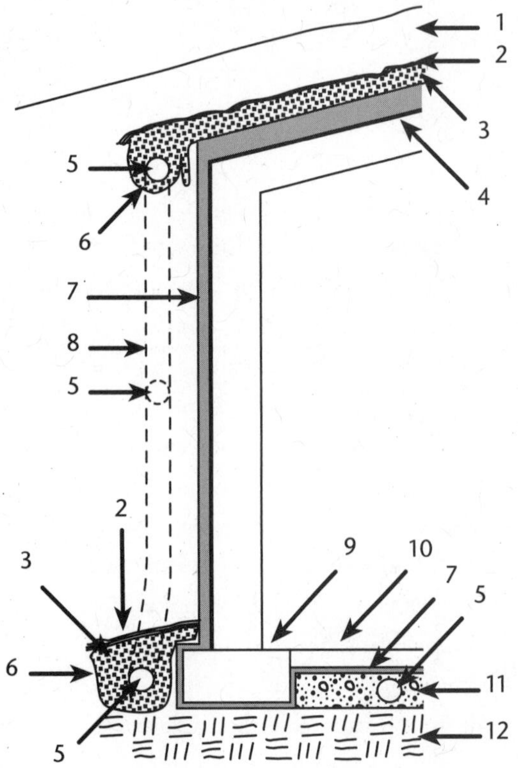

Fig. 7.13: 1. Earth. 2. Filtration mat; can be hay or straw. 3. Crushed stone. 4. Waterproofing membrane. 5. 4" perforated drain wrapped in a filtration sock. 6. 6-mil polyethylene to form underground "gutter." The top gutter should include a fold in the plastic as shown, to allow for earth settling. If you can get it, 10-mil polyethylene would be even better. 7. Extruded polystyrene insulation. 8. Vertically placed 4" non-perforated drains connect the horizontal perforated drains. T-junctions are available to make these connections. 9. Footing. 10. Concrete slab floor. 11. Compacted sand, gravel, or crushed stone. 12. Undisturbed subsoil or heavily compacted pad, as described in text. Note: As per text, composite drainage matting is an alternative to the crushed stone drainage layer on the roof, saving about 20 pounds per square foot on the roof load.

is it difficult to maneuver in the space, but it puts extreme vibratory pressures on the block wall.

Soils with poor percolation, such as clay or mixed clay/loam, should not be backfilled directly against the insulation. Water will not find its way to the footing drains, and will stand against the wall, putting unwanted hydrostatic pressure on the wall and the membrane. In this case, use a drainage matting made for the purpose, such as one of the several Enkadrain® products made by Colbond. (See Appendix B.) Enkadrain® consists of an intertwining nylon mesh, generally about a half-inch thick, protected on one or both sides by a filtration mat. The rolls are 39 inches (1 meter) wide by 100 feet long, with three-inch overlaps of filtration matting. The rolls are draped vertically on the wall and extend right over the footing drain.

With poor backfill – lots of clay for example – you have two choices. You can bring in good percolating backfill to place up against the wall, or you can use one of the many drainage mats made for the purpose. This will probably be a cost-driven decision. Remember that with bringing in backfill, the haulage is generally the largest part of the cash outlay, not the material. If you are some distance from the supply, haulage can be very expensive. We were fortunate at Log End Cave in two respects: (1) we owned the gravel pit where the backfill material was coming from (now the Earthwood site), and (2) the gravel pit was just one-half mile from Log End. A backhoe stayed at the gravel pit and loaded the trucks as they made the short journey back and forth.

DRAINAGE MATS (COMPOSITES)

There are now a variety of drainage products available for both sidewall applications and for use with living roofs. They vary slightly in composition from manufacturer to manufacturer, but they general work in the same way. Most drainage sheets come in 3- to 4-foot-wide rolls that are generally about 100 feet in length. The principle for all of these products is the same, in that they have a drainage core covered with a filtration mat. The products effectively place an uncompressible layer of between a quarter- and a half-inch against the insulation or the water-proofing membrane. A nylon or polypropylene filtration mat on the dirt side of the product stops

Fig. 7.14: From the lowest point of Log End Cave's footing drain – the southeast corner – the 4" perforated drain tubing continues along the back of the east side retaining wall, finally entering a soakaway.

Fig. 7.15: Jaki spreads hay or straw on the crushed stone surrounding the drainage tubing.

sands, silts, and other fines from entering the airspace. Some of the products have a smooth surface or a protection matting to go up against the wall, others do not. Enkadrain® from Colbond Corporation makes both kinds for different applications. A list of companies making these drainage products appears in Appendix B.

There are various means of installation on sidewalls. Cetco, for example, manufacturer of the Aquadrain® line of "subsurface drainage composites," says that the mats are installed vertically and fastened with "washer-head fasteners, general construction adhesive, two-sided tape, furring strips, or insulation pin anchors." Carlisle's Miradrain® composite,

according to the manufacturer, "should be attached with CCW Drain Grip adhesive or CCW SecureTape. Apply Drain Grip around the panel edge and in 4-inch ribbons on the back of the Miradrain® and on the corresponding surface of the CCW Membrane. After the Drain Grip is dry, mate the two surfaces together."

If you use Carlisle's Miraclay® bentonite-filled membrane over a concrete poured wall, you can actually nail the CCW Miradrain® right through the membrane to the concrete, using concrete nails and washers.

Colbond's Enkadrain® 3611R and 3615R "geocomposite" drainage sheets come in 100-foot rolls, 39 inches (1 meter) wide, and 0.45-inch thick. Sidewall application recommendations from the manufacturer are to hang the drainage mat from a "termination bar" at the top of the wall, which stops soils from entering the drainage space. In conversation with a Colbond representative, we agreed that an owner-builder could fasten a termination bar to an above-grade part of the wall and cover it with aluminum flashing to keep the dirt out. Such a home-made termination bar could be made of 2-by-2 pressure-treated wood. Alternatively, the top five inches of the polypropylene core can be removed from the top of a vertically applied sheet and the filtration fabric can be glued to the wall with construction adhesive. This also keeps the core free of dirt.

Colbond's literature does not speak of exterior rigid foam insulation. My inclination, on sidewall applications, is to stay with the rigid insulation against the membrane, and to apply the drainage sheets on the earth side of the insulation. With two-inch Styrofoam® sheets, for example, the drainage composite could be held to the Styrofoam® with 1½-inch roofing nails, until the backfill is installed. This way, water is drained away from the insulation, keeping it in better condition. The top of the drainage can be protected from dirt infiltration by one of the methods described in the previous paragraph.

In short, you will want to follow the manufacturer's installation instructions for their own product. Some of the data and installation sheets supplied are a bit cryptic, so do not be afraid to call the manufacturer with specific questions or for clarification.

ROOF DRAINAGE

Excellent drainage is just as important on the roof as on the sidewalls. I have used two entirely different drainage methods (crushed stone and composite drainage matting) and both work. But both methods also require the installation of a layer of 6-mil black polyethylene, which I personally feel is the best $50 to $80 that you will spend on the entire house.

Remember that we have created a roof pitch of between 1:12 and 2:12. Now, we want to use that pitch to our advantage to deliver the water quickly and easily to the drip edge. The 6-mil plastic is an inexpensive tough layer that provides a surface upon which the water can travel.

Without it, a lot of water will find its way between the rigid foam insulation and stand against the membrane. So, the 6-mil also serves to protect the Styrofoam® from water, too. We did not use it at Earthwood, but I wish we had. In December of 2003, we had a leak develop over a cordwood wall in the upstairs bathroom, the result of carpenter ants chewing upward and making a neat 1-inch by ¼-inch slot in our Bituthene®. Temperature-wise, a window of opportunity opened up on Christmas Day and I quickly tore off about 50 square feet of earth, crushed stone, and Styrofoam® to get to the leak and repair it. Despite being closed-celled, some of the Styrofoam® had taken on moisture. You could tell by its weight. I believe that if the insulation had been fully covered with the 6-mil plastic, the Styrofoam® would have stayed dry.

The plastic sheet is not intended to replace the waterproofing membrane. It is the base of the all-important drainage layer, a surface upon which the water rides to the edge of the roof. It might even get punctured here and there. But it carries 95 percent or more of the water away, taking a tremendous amount of pressure off of the membrane. I tell students that you want to think of the membrane as being like that well-known Maytag repairman of the TV commercials. The Maytag appliances are built so well that the poor bored repairman has to sit around the repair shop with nothing to do. That's the way you want your membrane to be: bored for lack of work.

Try to lay down the 6-mil polythene sheet in a single monolithic piece, which is not difficult to do with a Cave-type roof. With an Earthwood (or Stoneview) multifaceted roof, you will have to make some folds and incorporate some laps. That's okay. Just think of how shingles work when you make any laps and remember that the goal is to shed away *most* of the water, not every last drop. The membrane is there for a purpose, after all.

Crushed Stone Drainage

On most of our earth roofs, I have used two inches of #2 (roughly 1-inch) crushed stone, covered with three inches of loose straw or hay. The organic material keeps the soils from clogging the crushed stone layer. Eventually the straw mats down to a very thin stabilized filtration barrier, and the earth does stay on its side of the barrier and out of the stone. This is how septic drain fields filled with crushed stone are kept clean, incidentally. The advantage of crushed stone and straw is that you are using less petrochemical products. And, if you have a nearby source of stone, it might also be a little cheaper than purchasing drainage composites. The downsides are: (1) Two inches of crushed stone adds 20 pounds per square foot to the load, and (2) It is an awful lot of work to get these tons of stone up on the roof and distributed to a fairly consistent two-inch thickness.

Drainage Composites, or Drainage Mats

These products have already been described above in the part about sidewall drainage. They save you 20 pounds per square foot on the roof, and they are much faster and easier to install than the crushed stone. I used Enkadrain® 3615R drainage material on the living roof at Stoneview, our octagonal guesthouse, primarily because keeping load down was a very important consideration. As for installation, even though the 8-faceted roof involved cutting the Enkadrain® into trapezoidal pieces, the project was still much easier than the crushed stone method. On a Log End Cave type roof, the drainage would go on very quickly and easily. Figures 8.6 and 8.9 in the next chapter shows us installing the Enkadrain® 3615R on Stoneview, part of a photo essay that shows the entire living roof process from membrane to planting.

The drainage composite goes right down over the 6-mil black polythene, with the filtration mat upwards. When we cut the Enkadrain® into trapezoids with a razor blade knife, the edge of the mesh was sometimes a bit sharp and might have even punctured the plastic here and there. Next time, I will use the Enkadrain® 3811R product in this application, which has a second protection mat on the undersurface.

We found that the Enkadrain® tended to slide down the roof a little as we installed it. We were constantly pulling it back up, like loose-fitting trousers. After a little creative thinking, we found that we could "stitch" the bottom course of trapezoids together, one to the next, creating a kind of tension ring which prevented the sheets from doing any further sliding. We put a short (24-inch) length of 1-inch by 2-inch board under the place where the ends of two trapezoids overlapped, then tacked the sheets together with small nails that had large (1-inch-diameter) plastic washers under their heads. (See Figure 8.9 in Chapter 8) Two or three of these special nails into each little wooden stick was enough to stitch the ends together. I spoke with a tech rep at Colbond about the problem of slipping, and he suggested that double-sided construction tape would have

stopped the 3615R product from sliding. We also agreed that I might have been better off with the 3811R product, with fabric both sides of the drainage core. This second layer of fabric would protect the polyethyene, would probably slide less, and would work better with double-sided tape.

With two inches of crushed stone as the drainage layer, there is no great rush to get the earth on. The stone acts as ballast and protects the various layers below from UV deterioration. I have left roofs over the winter topped with crushed stone, installing the earth in the spring.

The living roof itself is the subject of the next chapter.

Chapter 8

THE LIVING ROOF

Earth-Sheltered Houses discusses techniques for building homes or outbuildings of low to moderate cost, which precludes massive earth roofs in excess of a foot thick. All of the techniques of the previous chapter are appropriate for earth roofs of 4 to 12 inches in thickness, although my personal view is that an 8-inch-thick earth roof is more than adequate to maintain a green cover, while providing the other advantages of earth roofs discussed in Chapter 1. With care in soil and plant selection, a 4-inch earth roof can be beautiful, ecological, and thermally effective. In fact, we'll open the chapter with a step-by-step photo essay showing the installation of just such a roof. This section also serves as a concise review of the roof layers described in Chapter 7.

STONEVIEW ROOF: A STEP-BY-STEP PHOTO ESSAY

The 5-by-10 and 6-by-10 rafters at Earthwood, on relatively short spans, carry a load of about 150 pounds per square foot, as per Chapter 1. The radial rafters at Earthwood's Stoneview guesthouse are only four-by-eights and the clear span is greater than at Earthwood itself. To make up the engineering shortfall, I needed to go with a lighter-weight living roof of about 115 pounds per square foot. This was accomplished by using 4 inches of topsoil instead of 6 inches (saving 18 PSF), and substituting Enkadrain® for the 2-inch crushed stone drainage layer, saving another 17 PSF. These pictures of installing Stoneview's roof (on pages 166–169) show clearly the steps taken to install a living roof in such a way that it won't leak.

THE EARTH ROOF

Once the drainage layer is on, you are free to install the earth. With composite drainage matting, it is good to do this as soon as possible, before it is lifted by the wind.

The thickness of the earth has been decided long ago at the design stage. Even a small shallow living roof like the one at Stoneview involves hauling a lot of soil by hand. There is an alternative, but you need to be careful. The boom of a backhoe can deliver 6 or 8 cubic feet of earth

Fig. 8.1 (right): You can't see the clear "surface conditioner" which Grace provides with the Bituthene® 4000 membrane, but it's there and has dried to a tacky feel. Large trapezoidal sheets were installed first at the perimeter, lapping 2.5" onto the aluminum drip edge. Bruce Kilgore (left) and Darin Roy set a small trapezoid of Bituthene® 4000 close to the center of the building. Darin smoothens it down while Bruce pulls the backing paper off. We lap 3" onto the previous sheet and 3" onto the exposed planking. Therefore, there is always a 6" lap on the cut ends of the sheet.

Fig. 8.2 (right): The author caulks the cut edge with Bituthene® Mastic.

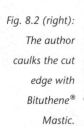

Fig. 8.3 (above): Jaki and Darin Roy feather the edge of the mastic with pointing knives, to approximate the factory edge of the Bituthene® membrane. The mastic stops the cut edge from raising up or "fish-mouthing."

Fig. 8.4 (left): Jaki installs the first layer of 1" Dow Styrofoam® extruded polystyrene over the membrane. Taping the joints simply holds them together until they are covered. We installed a second one-inch layer on top of the first. Using 2" thick sheets would have been quicker, but we bought these 1" sheets at a very low price and saved a lot of money.

Fig. 8.5 (above): A layer of 6-mil black polyethylene serves as the base of the drainage layer and is installed over the insulation. This is tricky to do on an octagon roof, as the sheet needs to be folded over the saddles between facets, but it is an easy job on a Cave-type roof. We were careful to give plenty of overlap from sheet to sheet (at least 18") and paid attention to the shingle principal when making our overlaps.

Fig. 8.6 (below): Trapezoidal sheets of Enkadrain® 3615R composite drainage are placed over the plastic, with the filtration mat uppermost. No, we haven't forgotten the plastic sheet. You can see it poking out here and there. A large centerpiece of plastic will be tucked under the Enkadrain® before it is covered with earth. Heavy wooden blocks keep everything in place until the earth goes on.

Fig. 8.7 (left): Installing a woodstove chimney. Any projection cut through the membrane has to be treated very carefully, with both waterproofing and drainage. As our Metalbestos® stovepipe has an 8" outside diameter, we made a hole of 12" in diameter through the Bituthene®-covered planking. A reciprocating saw works well for cutting the hole. This way, the chimney has the required 2" clearance from combustible material. The Roof Support Package (Metalbestos® part RSP) unit shown has two adjustable metal flanges which can be screwed to the planking, allowing the installer to plumb the stovepipe itself to a true vertical.

Fig. 8.8: After the stovepipe is plumbed, the aluminum Metalbestos® flashing cone for shallow-pitched roofs is nailed in place. Using 5″ strips of Bituthene®, we covered the roofing nails and the join where the aluminum overlaps the roofing membrane. Bituthene® mastic around the edges completes the waterproofing. The storm collar (Metalbestos® part SC) keeps water from dripping down the stovepipe and into the flashing cone area.

Fig. 8.9: Over the Enkadrain®, we installed 4-by-4 pressure-treated timbers to retain the earth. (An alternative method of retaining earth, involving cut sods, is discussed elsewhere.) Inexpensive galvanized truss plates tie the eight retaining timbers together to create a ring beam with good tensile strength. At Stoneview, the Enkadrain® tended to slide down the plastic, even with the relatively shallow 1:12 pitch roof. As described in the text, we stitched the bottom ring of trapezoidal sheets together with 24″ pieces of 1-by-2 wooden scrap. The picture shows a couple of large plastic washers under the nail heads, holding two adjacent pieces of Enkadrain® lapped and firmly attached to the wooden piece, out of sight below the lower sheet. Once this ring of Enkadrain® was stitched together all around the building, it stayed in place and further sheets were simply laid down over each other, covering previous sheets with the 3″ filtration mat factory-installed on one edge of the roll.

Fig. 8.10: The author hauled three quarters of the roof's topsoil up the stairway in 5-gallon buckets. Bruce Kilgore helped with the other quarter of the estimated 400 bucketloads.

Fig. 8.11: The author uses the back of a rake to spread the soil to a consistent loose depth of 5½". Twenty-four 6-by-6-by-8-inch blocks (actually 5½" thick) served as depth gauges. The soil slumps down to the 3½" deep retaining timber. After a winter's snowload, the loose soil compressed to a compacted depth of about four inches.

Fig. 8.12: Drainage is the better part of waterproofing! Any projection through the membrane is a potential leak. I boxed around the stovepipe with pressure-treated 1-by-6 material and placed crushed stone over the Enkadrain®. Water is quickly carried away to the edge of the building.

Fig, 8.13: Jaki plants a couple of dozen clumps of chives between 100 sedum plants (protected by mulch) on the Stoneview roof. Sedum is a succulent plant with the ability to survive prolonged drought conditions. Chives also do well in dry conditions.

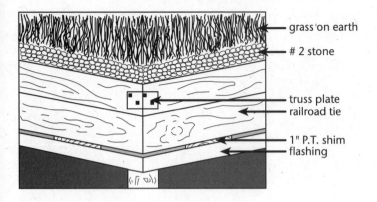

grass on earth

2 stone

truss plate
railroad tie

1" P.T. shim
flashing

Fig. 8.14:
Construction
details when
using rail ties or
landscaping
timbers at the
roof's edge.

at a time. That's a lot of 5-gallon buckets. But it needs to be placed carefully on a part of the roof where the load is transferred directly by compression to the foundation. Had I used a backhoe at Stoneview, for example (and it crossed my mind), I would have had the operator place the soil right over the very center of the building, which is heavily overbuilt. (Sixteen rafters come together over the capital, supported in turn by a 15-inch-diameter post capable of taking over 40 tons on compression.) Have two or three people on the roof, armed with shovels, rakes, and buckets, to distribute the soil as it is placed on the roof. An earth roof can go on quite quickly with some organization and strong willing bodies.

With earth roofs, in the 4- to 8-inch range (settled thickness), use soils that will retain moisture. Sandy soils will transmit water to the edge of the building quite quickly, making it difficult to maintain a green cover without constant watering. Earth roofs should be low maintenance and should not require watering,

which is why we are using sedum and other drought-hardy plants at Stoneview.

RETAINING EARTH AT THE EDGE WITH TIMBERS

With a Cave-type situation, the earth roof meets seamlessly with the earth berm. But on the north and south parts of the roof, or with freestanding earth roofs like those at Earthwood and Stoneview, you need some sort of retaining edge to keep soil on the roof. At Earthwood, we got a great deal on railroad ties in excellent condition, just $100 for twenty ... delivered! They came from a local rail yard where a section of track was being removed. You could use 6-by-6, 6-by-8, or even 8-by-8 pressure treated landscaping timbers for the purpose, but these can be quite expensive. Four-by-fours are a lot cheaper and were thick enough for the lightweight living roof at Stoneview.

Figure 8-14 shows how we join two rail ties (or landscaping timbers) on an 8-sided or 16-sided roof. (See also Figure 8.9.) A critical detail is to keep these timbers off of the roof substrate, so that an ice dam does not form behind them. Initially, we made the mistake of placing the rail ties right down on the inch of Styrofoam® that we used as a protection board for the Bituthene®. Two or three years later, about 1984, we had our first leak, which was also our last until December 2004, already mentioned. It rained for two days in a row, but air temperatures were below freezing. It can happen. Water went through the crushed stone drainage layer to the overhang, where it would freeze because it was dammed behind the rail ties, the classic "ice dam" situation.

Water has no choice but to back up onto the roof. When drainage can't work, the membrane is severely taxed. We had a leak in our bedroom. In the spring, we tore up a section of the roof and found a neat little hole in the Bituthene®, created by carpenter ants from below. We repaired the leak, but then levered up every one of the 200-pound rail ties, and slipped three or four 1-inch pieces of pressure-treated boards as shims. This was not a big enough gap to allow the crushed stone through, but was sufficient to allow the water to get to the drip edge. At Stoneview, the four-by-four retaining timbers actually sit on the Enkadrain®, so that the water can find its way under the timbers and to the drip edge.

In 2002, we added an upstairs sunroom to Earthwood, which broke the continuous tensile ring of 16 rail ties, fastened together with galvanized truss plates. At that time, I found another use for the rail ties, and replaced them with moss sods, as described below.

RETAINING EARTH WITH SODS

Taking a leaf from Mac Wells' well-worn book, we used moss sods to retain the earth on the freestanding earth roofs on the Earthwood library and office buildings, which can be seen in the color section. We have also used grass sods successfully. We prefer the natural appearance of the moss (or grass) sods on the edge, and they save

Fig. 8.15:
Sods are cut
with a sharp
square spade.

Fig. 8.16: The author installs a moss sod on the edge of the office roof.

a lot of money over wooden edging methods. We are fortunate to have a good source of moss sods growing in sandy soils at Earthwood, although any well-knitted grass sods will work as well. In Figure 8.15, I am cutting sods with my sharp square spade. I like to cut them about 5 inches wide, 5 inches high, and 10 or 12 inches long.

In Figure 8.16, I am installing a sod on the office roof.

On a shallow roof, the sods stay put and knit together nicely. It is a very rare occasion, usually during drought, when a small bit of earth falls off the edge of the roof, rarely anything larger than a baseball. Moss goes dormant during a drought, but bursts back into beautiful life with a good soaking rain. We have three or four different varieties growing on our roofs.

Figure 8.17 shows the detailing with moss or grass sods as a retaining edge. The sods go right down on the 6-mil black plastic which covers the Styrofoam® protection board.

Another successful detail, not seen in the diagram but evident in Figure 8.16, and in 8.18 below, is to install a 10-inch long piece of half-pipe every three or four feet around the edge of the building. This drainage aid, ripped from a piece of 3-inch ABS or PVC pipe, extends under the sods and meets with the crushed stone drainage layer. During the winter, we often see little icicles forming all along the drip edge around the office, but large icicles form where the half-pipes are located, showing that they provide a positive relief to hydrostatic build-up behind the sods.

With composite drainage matting, like Enkadrain®, I take the drainage sheets right to the edge and put the sods – or retaining timbers – right on top of it. The half-pipes are not needed in this case.

LOG END CAVE: THE EARTH ROOF

We wanted an "instant grass roof" at the Cave in the fall, one that wouldn't erode over the winter. In June, we tilled, de-stoned, and planted a section of the front field to timothy and rye. In October, we cut a few sods from the sod field and found that 2 or 3 inches of soil came up with each piece. To get our desired 6-inch thickness, we spread 4 inches of topsoil over the entire roof, and covered it with plenty of hay mulch and pine boughs for erosion protection. Finally, long heavy sticks kept the pine boughs from blowing away. A friend experienced in golf course maintenance advised me that it was too late in the year to install the sod on the roof, but I did do 20 square feet of sod in one spot as a test. My friend was right. We sodded the rest of the roof in early June of the following year, and it was flourishing and in need of a second mowing by July 1, while the test patch was still in bad shape.

Speaking of mowing, the earth roof at Log End was really an extension of our lawn, and mowing seemed appropriate for esthetics. At Earthwood, we mowed for a couple of years, but then my environmentally-minded son, Rohan, pointed out that every hour that we run the heavily polluting two-stroke lawnmower puts as

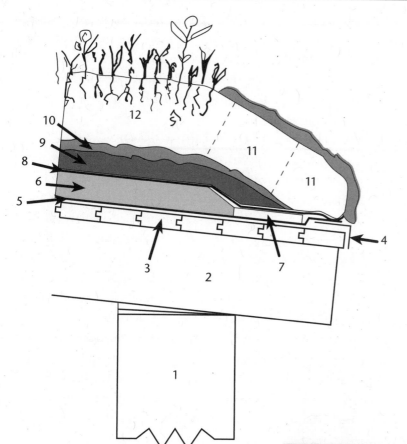

Roofing detail for a free-standing earth roof using moss sods to retain the earth.

Key:

1. Above-grade wall.

2. Heavy wooden rafter.

3. 2" × 6" T & G planking.

4. Aluminum flashing as drip edge.

5. W.R.Grace Bituthene® 4000 or equal membrane.

6. 4" to 5" rigid-foam insulation

7. 1" rigid foam or half-inch fibre board to protect membrane.

8. 6-mil black polyethylene.

9. 2" of #2 crushed stone drainage layer.

10. Hay or straw filtration mat.

11. Moss or grass sods cut from sandy soil, retain the earth at the edges.

12. 7" to 8" topsoil, planted.

much pollution into the atmosphere as driving 300 miles in a car. We have not mowed an earth roof since, and prefer the natural look. Grasses left long also withstand drought much better than mowed roofs.

Sodding was a lot of hard work. Three of us — one cutting and two hauling and laying — applied the sod onto the topsoil in two days. As we also covered the east and west berms, this amounted to about 1,600 square feet. We found that sods larger

Fig. 8.18: Moss sods drain well, retain the earth and look natural. Note "half-pipe" for drainage.

than ten inches square would break up in transport. Sods should be cut damp, not soaking, so that they hold together. It doesn't take sods long to knit together and form an instant green roof in their new location.

At Log End Cave, we accidentally discovered an easy way to start a green roof. Our mulching hay from the previous October was full of seeds, and the roof was becoming quite green on its own before we transferred the sods. Had we known, we could have applied the whole six inches of soil in the fall and let nature take its course in the spring.

We only cut sods now for instant edge retention.

EARTHWOOD: THE EARTH ROOF

We let the Earthwood roof sit over the winter with just the 2-inch crushed stone drainage layer on top, sufficient ballast against the wind. We installed the earth in August of 1982, at a time when the backhoe had returned to the site to finish landscaping.

First, we covered the crushed stone with three inches of loose hay, which forms a natural filtration mat under the earth, keeping the crushed stone drainage layer clean.

The "topsoil" was actually silty material that came from a stream's floodplain on a local farm. The soil is free of stones and has rather poor percolation, perfect for retaining moisture. Ed Garrow carefully used the loader end of the backhoe to place the soil on the roof. Several of us spread the soil with shovels, 5-gallon buckets and rakes.

Learning from Log End, we planted rye in early September, and, thanks to an idyllic autumn, had grass coming through a thin mulch layer in weeks. By November 1st, the grass was thick, green, and lush.

OTHER LIVING ROOFS

All earth roofs should wind up as living roofs, but all living roofs do not necessarily require a layer of earth. During the 1990s, we built three small outbuildings, all in the woods, with roof areas of 100 square feet up to 320 square feet. Influenced by work which others had done in the natural building field, we experimented with hay bale roofs. On the first one, a guest house called La Casita, we put up all the layers of waterproofing membrane, insulation, and drainage, as described in Chapter 7, but, instead of sods or topsoil, we finished the roof surface with full bales of hay, packed tight against one another and tied together around the perimeter with baling twine. We never cast any seeds, but, within a year, all sorts of green things were growing up there. No doubt there were lots of grass (and other) seeds in the bales, and birds and wind probably brought other seeds. After a couple of years, the bales had compressed to a third of their original thickness and the roof was positively lush. Moss took a foothold and poplar trees got established. After six years, I pulled all of the poplar trees, fearing that their roots might create havoc with the membrane. Now, in the spring of 2005, ten years after the hay bales were installed, the roof is still alive, with mosses and wild blackberries mostly, and the bales have compressed and composted with years of leaves to form a three-inch layer of rich, black, lightweight humus. A white birch tree has established itself. I hate to pull it out.

The Straw Bale guesthouse and the composting toilet were done at about the same time, around 1997. Again, we covered the buildings with hay bales, straw being uncommon in our area, but we also placed a good inch of topsoil on top of the bales, and seeded the roofs to grass. These roofs established themselves even faster than at La Casita, and became green and lush within weeks. One year, the Straw Bale guesthouse was covered with "pigweed," also called "Lamb's Quarters," an edible plant, good in salads or cooked like spinach. These roofs seem to have maintained the hay thickness better than the roof on La Casita. They are not as old, it is true, but I expect that the topsoil has something to do with it as well.

All three living hay-based roofs have shown great resiliency against drought conditions over the years, probably because they are in very shady areas, and protected from the worst of the wind. There seems to be an irregular turnover of vegetation, but that's okay. I reckon that whatever is growing up there is the right thing. These are wild roofs, not a great deal different in appearance from the forest floor, and so they fit in well with the habitat.

I wish we had ready access to straw. I had to go 150 miles to get bales to build the Straw Bale guesthouse. My thinking is that bales of straw, topped with an inch of topsoil for seed starting, would make an excellent living roof, particularly in a shady or partially shaded area. The straw bales would not decompose as fast as hay bales, but would provide a good medium for root growth.

WHAT TO GROW ON THE ROOF ...

... is a function of roof thickness, the growing medium, climate, and whether the site is sunny, partly sunny, or mostly shady. We've been happy with our living roofs, so, in the spirit of writing about our own experience, and not what we have heard second-hand from others, here it is:

Our major earth roofs at Log End and Earthwood had compacted earth layers of 6 to 7 inches. Various grasses (timothy, rye, even tough local "Johnson" grass) and clover have provided most of the vegetation on these roofs. I have described how we sodded the Cave roof and planted grass on the Earthwood house.

The 350 square-foot office roof, with a similar thickness of earth, presented a slightly different situation. In the fall, we planted what was supposed to have been "annual rye;" that is, it was supposed to survive as a green cover until spring, and then die off. Well, there must have been a percentage of perennial grasses mixed in with the seed. Most, but not all, of the nice green grass cover did die off. Our intent was to plant a temporary green cover until we could grow a wildflower roof in the spring. Well, we did have great success with the seeded wildflowers ... for about a year. We broadcast the contents of a can of wildflower seeds onto the almost (but not quite) barren roof in the spring. The label on the can claimed "26 varieties" of wildflower seed. And the results were glorious, as you can see in the color section. But, of the 20-odd species that grew the first year, only three or four returned the second year because the "annual rye" took over the roof. By the third year, only a few odd wildflowers popped up here and there, and now they are rare. Grasses seem to trump wildflowers, except, perhaps, in the wild. If you want wildflowers, plant them without grass. Eventually, though, grasses will probably arrive spontaneously on the roof and choke out the flowers.

We were sorry that the office roof didn't look like that wonderful first summer in subsequent years, but, you know, it still looks pretty darned good, with its moss edge and the variety of grasses and other wild plants, some of which flower, most of which don't. There is no such thing as a weed in the wild, at least not insofar as they don't belong there. (The exceptions to this idealistic comment, I suppose, are the "invasive species" which choke out native plants, vegetative varmints like kudzu in the South, and purple loosestrife in the North.) The office roof does change character with the seasons and the amount of rain. We do not mow it and it almost always looks good.

The library roof has never been a great success, except that it is at least alive and not dead. But (until recently) it was never a vibrant roof, except for mosses, which seemed to be happy up there. We suspected the soil was sandier and lacking in nutrient. Combined with its smaller roof area, the net effect is that the library roof was more prone to damage from drought. In early May of 2005, we attempted to rejuvenate the roof by planting some sedum and other plants here and there, and I have added considerable nutrient by raking in some well-composted "humanure" from the composting toilet. A month later, the roof was looking better than it had in years.

It is important to note, however, that while not quite as pretty as the other roofs, the library roof was never ugly, either, and, esthetics aside, still accomplished the other earth roof goals and purposes.

A certain amount of shade seems to be a plus for earth roofs, making them more proof against drought. Our small sauna roof, just 112 square feet, has a lush cover of various grasses. The soil is no thicker than other roofs, but the building is shaded by the house for three or four hours each summer day, a break from the sun not enjoyed by the library or office roofs. As I write, the lushest roof at Earthwood happens to be the little (28 square feet) roof on the former Littlewood playhouse, now a kindling storage depot. This roof gets more shade than the others, morning and evening, but also gets full sun for several hours when the sun is at its highest.

The guesthouses and outhouse in the woods, with their hay bale living roofs, we leave completely to nature, save pulling trees out every few years.

PLANTING A LIGHTER-WEIGHT EARTH ROOF

Nigel Dunnett and Noël Kingsbury, in their fine *Planting Green Roofs and Living Walls* (see Bibliography), comment that: "To achieve its function, rooftop vegetation must be able to:

1. Cover and anchor the substrate surface within a reasonable time after planting;

2. Form a self-repairing mat, so that new growth will be able to fill any areas that become damaged, for example, through drought;

3. Take up and transpire the volumes of water that is planned for the water balance of the structure; and

4. Survive the climatic conditions prevailing on the rooftop, with particular attention to cold-hardiness and drought tolerance; worst-case weather scenarios should be assumed."

All the literature – and our conversations with those with more experience than ourselves – points to sedum as the plant of choice on a lightweight (4-inch or less) earth roof. In fact, sedum has done well with as little as an inch of soil. Why? What is this stuff? I'd never heard of it until one of my students told me about it in 2004. Later that year, the same student gave us a tray of various sedums – there are hundreds of varieties – to experiment with, and we have now planted them on the library roof.

Sedum is the name for a genus of plants characterized by having a moisture-retaining leaf system, like succulents or cacti. Dunnett and Kingsbury enthuse: "Sedums have ... become the bedrock of shallow-substrate roof-greening systems for their drought tolerance, year-round good looks, ease of propagation, and suitability for shallow substrates."

We live in a very cold climate, but we knew from speaking with Chris Dancey, author of the case study that closes this chapter, that there are sedum varieties tolerant to severe cold. Her own beautiful living roof features sedums in Ontario, a climate similar to our own in northern New York. For the 4-inch-thick Stoneview roof, Jaki ordered from plant catalogs that she has dealt with in the past and selected two varieties: Dragon's Blood (*S. spurium*), a brilliant red-flowered sedum from Gurney's, and Improved Golden Sedum (*S. kamtschaticum*) from Spring Hill Nurseries. Both varieties were recommended for cold climates. She ordered about 100 plants in all. She also ordered 24 Vinca Minor plants, for variety. Vinca is not a sedum, but an evergreen foliage "super in sun or shade and in any soil," according to Gurney's. It will end up about 6 inches high in a tight mat. "Royal purple flowers bloom in spring."

Grasses, fescues, and sedges can be grown with a soil thickness of 4 inches or more. If you are looking for a lawn type of situation, go with the regular species for that purpose: rye, Kentucky bluegrass, fescue, etc. If you want wildflowers to coexist with your grasses, you might want to select shorter ornamental grasses. Although written from a British point of view, *Planting Green Roofs and Living Walls* does give a lot of information on grasses and other potential green covers for shallow roofs. An important consideration, however, will be suitability and hardiness for your climate, so it is a good idea to get advice from landscape gardeners and suppliers in your area. Let them know that you have shallow soils and need something that is drought-resistant as well as cold-hardy.

Jaki had a huge clump of chives growing in one of the raised beds. She read in *Planting Green Roofs and Living Walls* that chives (*Allium schoenoprasum*) "survive well on green roofs under both wet and dry conditions," so we hauled the clump up onto the Stoneview roof and she broke it into a couple of dozen small clumps, which, hopefully, will spread between the sedums. We have also planted a variety of wildflower seeds while there is little or no grass up there, to provide some vegetation and color until the sedums take hold and spread. The container lists 24 varieties, including various poppies, primroses, lupine, daisies, and baby blue eyes, a mixture of annuals and perennials.

To avoid introducing an invasive species, use only local native plants or ones that have been cultivated in your area for a long time without causing problems. Plant suppliers should know which plants are safe and should not be selling you anything which can cause environmental damage, but, as Dunnett and Kingsbury point out, "There are as yet no horticultural industry protocols on the risk assessment of introduced species."

Avoid strange and exotic species. If in doubt, don't.

MAINTENANCE OF LIVING ROOFS

We have done very little to maintain our several earth roofs. I'm not advocating this approach, you understand. But here are some comments, for what they are worth:

I don't think that a roof that requires watering is a well-designed roof. And mowing a roof doesn't make much sense, either.

Because we don't mow, there is sometimes a lot of dead chaff and stubble on the roof that seems to choke off growth. I have, on three or four occasions, burned off this dead material. I choose my time carefully, late in the day, when it is cool and damp and never at a time of fire danger to the surrounding grasses or forest. Each time I have burned the roof off, it has bounced back lush and green. My neighbors think I'm nuts, but with a megalithic stone circle in the front yard since 1987, I doubt if burning off the roof has greatly altered their opinion.

The other "maintenance" I have performed is to remove poplar and birch trees, which establish themselves. A shallow roof is not the right place for trees, shrubs, or root crops. (Leafy vegetables, however, are an option for those who are otherwise short of garden space.)

Where we have been remiss is in not feeding the roofs once in a while. By not mowing, the roof achieves equilibrium similar to a natural habitat, but I am sure that a little feeding once in a while, with organic fertilizers such as bone meal, and with ash from the woodstove, will help plants to thrive. And now I've found a good safe use for our humanure.

LIVING ROOFS ON A COMMERCIAL SCALE

Living roofs are big business in Europe, particularly in Germany. For the most part, the projects are commercial, municipal or industrial. But the reasons for choosing living roofs on commercial buildings are very similar to the reasons listed in Chapter 1 for homes and small projects. An additional advantage to living roofs in cities is that they can effectively change the micro-climate for the better: every black tarscape converted to a green, oxygenating surface has got to help the overheating problems prevalent in urban areas. Another reason for green roofs cited in the commercial literature is that they last much longer than conventional roofs.

The idea is catching on in the United States, too. Ford Motor Co. made the headlines in 2002 with their nearly 10-acre earth roof renovation of their Dearborn automobile plant. An American firm, Colbond (makers of Enkadrain®, already mentioned) had an involvement with the Ford project, as well as the 320,000 square-foot Millenium Park in Chicago and the 180,000 square-foot International Plaza in Atlanta. Of the Ford project, it is said that the roof cost twice as much as conventional roofing, but will last twice as long. Ford's executives figured that the positive PR of building "the world's largest ecologically inspired roof" was worth the cost.

Colbond manufactures a "root reinforcement matrix" called Enkamat® R²M 7010, made to "permanently anchor plant roofs on sloped roofs or in high wind conditions. ... As the roots grow they become entwined within the Enkamat®, making an extremely stable cover. Its tough root reinforcing system anchors vegetation and provides a holding cavity for the growing medium."

Then, taking convenience a step further, Colbond provides pre-planted Enkaroof® VM, which the company describes as "an Enkamat® core with a non-woven fabric attached to the bottom side designed for pre-vegetated mat applications. The mat is filled with the growing medium and plants are grown directly in the mat structure. ... The fabric holds soil medium and vegetation in place while the flexible matting is re-rolled and shipped to the installation location."

Other companies have products designed to do the same sort of thing: provide a fast, proven, lightweight living roof. Appendix B lists several of these companies, and some excellent websites that will put you in touch with what is happening in the – pardon me – fast-growing Green Roof industry.

Are these commercially oriented systems worth their cost? Depends on the size of the

Fig. 8.19: Flowering sedum makes an attractive summertime cover.

growing medium, can keep the additional weight of the living roof layer down to just 10 pounds per square foot, considerably better than the 4-inch saturated soil load of 40 pounds per square foot at our "lightweight" living roof at Stoneview guesthouse. These systems could pay for themselves on reduced structural costs, or might provide a green-roof alternative where spans greater than 12 feet are needed.

I thought it would be good to give the reader a break from my voice, and learn a little about green roofs from my friend Chris Dancey, who has taken the medium to a more esthetic level in Ontario. So I'll give Chris the last word in this chapter.

project and your budget. The techniques accented in this book are directed towards the owner-builder on a low to moderate budget, but there may be circumstances where modern-day living roof technology might be helpful, so it is good to be aware of it. For example, some of the sedum mat systems, with just 2 cm (less than an inch) of

CHRIS DANCEY: OUR LIVING ROOF IN ONTARIO

Our living roof brings us joy year round, through its beauty and ever-changing nature. In five years it has progressed from small patches of succulents, mostly sedum, to a dense mat of ever-changing color. Even during winter, it is elegant in a cloak of white, with a hint of the life beneath.

Two reasons motivated us to construct and care for a living roof. The first was simply the desire to have a beautiful roof and the second was to demonstrate for others that alternative roofs are a viable choice.

In urban areas the motivation may be different, since living roofs can help control pollution and provide garden space. Whatever the motivation, remember that having this style of

Fig. 8.20: This view of the Dancey home in Ontario was taken in January, 2004.

living roof is not as maintenance-free as most roofing systems. Now that my roof is established, I do roughly two hours of weeding and maintenance each month, for about eight months each year.

On the positive side, a well-designed living roof can extend the life of a waterproof membrane by many years, even decades, as the soil and plants protect the membrane from ultraviolet rays and extreme temperatures. Our waterproof membrane is a two-layer product from Soprema (listed in Appendix B), which was designed for this purpose.

We live in southwest Ontario, Canada, about ten miles north of Lake Erie. Our latitude of 43 degrees is similar to Rome, Italy, and southern Oregon. The Great Lakes have a huge effect on our weather. Summer can be hot and humid, or dry with weeks of drought. Cold, snow-laden winds are common in winter, but we can also have mild winters with little snow. So far our roof has thrived in the varied conditions, mostly because of the diverse selection of plants. For me, this is part of the beauty of my roof.

The building below our living roof is heated roughly six months each year. The roof is insulated, so the living roof and snow cover are bonus insulation and would never be adequate on their own in a cold climate.

Summer is when you really notice the positive insulation value of a living roof. The roof surface doesn't heat up and the plants and soil have a cooling effect. Even with two large roof windows, the building is always comfortable inside.

Chris Dancey

The composition of our soil is equal parts sand:clay:soil. The sand helps with drainage. The clay holds rainwater and then releases it slowly. The quality garden soil provides nutrient and is a good growing medium. We bought this mixture from a landscaping company that was able to create a custom blend for us.

When we put the soil on the roof, it was about 6 inches deep. Over time, we knew the soil would settle to about 4 inches in depth. Walking on the roof during planting and maintenance has also compressed some areas. In 2003, I took a bit more soil up to fill in the depressions. The original soil and this new soil carried lots of seed for plants I didn't want on my roof. Until my selected plants were able to provide an effective ground cover, I had to do a fair bit of weeding.

The slope on our roof is 18 degrees, which is a 4:12 pitch. About two days after our roof was

Fig. 8.21: The Danceys collect excess water from the roof for use in their garden.

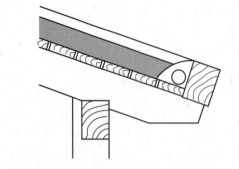

*Fig. 8.22:
Drainage detail
at the bottom of
the slope.*

*Fig. 8.23:
Photograph of
the drainage
detail at the
Dancey home.*

Chris Dancey

easy to establish, and they stabilize the soil. The dense, leafy growth shields the soil from the drying effects of the sun and wind. Condensation is adequate to keep them alive during a drought.

Profuse flowering begins with the self-seeding violas and violets in the spring. As the days become hotter and the soil dryer, these plants die back, but not until they have set seed for the next cool damp spell.

Before choosing the plants for our roof, I looked at several roofs in Germany and many published photos. I knew the style and plants I chose would require more maintenance than a sod or 'wild' style, but I've never regretted the choice. For me, the ratio of work to pleasure is excellent.

Some Technical Details

A 3-inch-diameter big 'O' tube that is in a fabric filter runs the length of the roof on both sides. Pea gravel, which acts as a coarse filter, surrounds the big 'O' tube. A 6-inch-by-9-inch oak timber forms the edge. The whole edge of the roof is sloped toward the drain outlet above the rain barrels. The entire wood surface of the roof is protected from moisture by the Soprema membrane. Metal flashing protects the upper surface of the timber.

When the soil is dry, it absorbs and holds the rainwater. Once the soil is saturated, the drainage system must be able to carry the excess water off the roof. A wet roof is also extremely heavy, so the structure must be engineered for the weight of the

planted, we experienced torrential rain for several hours. I was afraid to look, thinking that the soil and plants might now be on the ground below. When I finally got up the courage to inspect, I found that the soil had settled, but there wasn't even a sign of run-off.

The plants I chose don't need nutrient-rich soil. The majority are drought resistant, succulent varieties that are used in poor soil conditions and rock gardens. They are low growing and send out roots on their new branches, which makes them

wet soil load. My husband, Wil, designed and built the pine timber-frame structure and he also did training with Soprema to enable him to do a professional installation of the membrane.

Two rain barrels catch the run-off. A garden hose hook-up allows me to use gravity to water the garden below. If the rain barrel reaches the overflow level, a second outlet channels this water into a vertical tube beside the building. Through a system of elbow and T-joints, the water is moved under the deck to the garden. The water is then distributed along the length of this garden, through a perforated tube – also known as *soaker hose* — that was buried a few inches under the soil, before I planted the garden.

During the winter, the rain barrels must be removed. We attach a big 'O' tube to the stainless steel spout and this handles any run-off from melting snow. Again, the water runs into the garden below.

Chapter 9

FINISHING THE EXTERIOR

To repeat an important point from Chapter 1: All houses are an imposition on nature, but the house with the potential for the least negative impact is the earth-sheltered house. Better than with any surface structure, we can return this little piece of the planet back to living oxygenating greenscape, instead of heat-producing deadscape. But "potential" and "we can return" are conditional terms. Sadly, many owner-builders take their project through to the point that the building becomes livable, and leave it at that, sometimes indefinitely. We are all familiar with the stereotypical tarpaper-covered buildings. Some people deliberately leave projects at this sorry stage in order to keep their valuations (and, therefore, their property tax burden) down to a minimum. But if the house is visible to all, then the only word I can think of to describe a "deliberately unfinished" strategy is: crass. With earth-sheltered housing, crassness is compounded by sadness, because the house is so close to realizing its unique potential for environmental harmony, and yet falls short. Some people never

really finish the project, and take comfort in the Chinese proverb: "Man who finish house, die."

Finish the house … inside (next chapter) and out. If you want to satisfy the proverb and live forever, leave some little detail undone.

And, with earth-sheltering, there are some very practical advantages for completing the outside work. Retaining walls are needed to keep the side berms from collapsing around to the front of the building, for example, and the vegetation acts as an effective erosion control.

Bracing before Backfilling

Once the walls are covered with waterproofing, insulation, and drainage, as described in Chapter 7, and the footing drains are in, we are almost, but not quite, ready to backfill the sidewalls. It remains to brace the structure against the lateral load, including concentrated and momentary loads which might occur during backfilling, and never again – such as a boulder rolling down the incline, or over-tamping the backfill with the backhoe bucket. Large rocks need to be watched

185

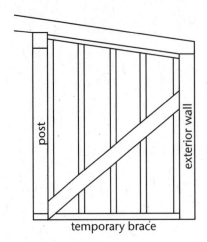

temporary brace

for very carefully, and not allowed to smash into the wall, and tamping must be gentle and careful. An experienced operator is the key here.

Even so, bracing the sidewalls with the internal house skeleton is imperative, and can more than double their resistance against the earth load. At Log End Cave, having the roof rafters in place gave a lot of support at the top of the east and west walls, greatly increasing their strength in this regard. But on the longer (35-foot) north wall, we felt that additional temporary bracing was necessary before backfilling, and we integrated this bracing with the floor plan, as seen back in Figure 5.9.

The bracing itself, seen in Figure 9.1, above, was easy to do, because we'd already framed most of the internal walls with 2-by-4 studs, and all that we had to do was to nail a diagonal brace to the framing for support, as per the illustration. The base plate of the 2-by-4-studded frame is nailed to

the floor with concrete nails, also called masonry nails. We braced all around the home's internal perimeter, wherever an internal wall met the block wall.

At the two-story round Earthwood home, it was necessary to do a partial backfill on the northern hemisphere, so that we had a place to stand to continue the blockwork on the second story. As the inner surface of the 16-inch-wide curved block wall is on compression – and the buttresses, already described, stop the building from moving south – we did not worry about internal bracing. The radial floor joist system was in place, however, before the partial backfilling. Although I use the term "partial backfilling," it is a significant amount of the total load, as the lower story is much more heavily earth-sheltered than the second story.

RETAINING WALLS

Almost every earth shelter will require some kind of retaining wall to make the transition from below grade (or bermed) space and the non-earth-sheltered portions of the home. And, usually the building of the retaining walls is most easily accomplished as a simultaneous operation with backfilling. Backfilling helps support the retaining walls, and vice-versa.

At both Log End Cave and Earthwood, I have used large stones to build the retaining walls. In each case, the stones were already on site. Not everyone has large stones available, and there are other retaining wall options, which I will discuss

On "Soakaways" or Dry Wells

At both Log End Cave and at Earthwood, we ran footing drains into a large stone-filled soakaway or "dry well," and this worked for us, but in areas of poor percolation, the soakaway might become full and no longer function. Therefore, taking the drain out to "daylight" is preferable.

However, if you have good percolation and a low water table, a soakaway might work for you, as it has for us. If you choose that route, be sure to slightly mound the top of the soakaway to carry surface water away from that area more easily. Fill the large hole with rocks from a stone wall, and cover the stones with flakes of hay or straw as a filtration mat, keeping soil out of the soakaway. The worst case would be if the soils above the soakaway settle into a kind of "dish."

Personally, I think we have been fortunate with our soakaways. I would only advise their use if (1) there is no other option and (2) if a 5-foot-deep test hole reveals that your soakaway will not cut into the water table.

below, but large stones are a great choice if you've got them, for two reasons. First, the spaces between them relieve hydrostatic pressure. Second, large stones tend to stay where they are placed. Also, the same heavy equipment used for backfilling, particularly a backhoe, is useful for building a megalithic wall.

Good drainage behind retaining walls is every bit as important as against the side of the home. The most common causes of retaining wall failure is bending or tipping against hydrostatic and frost pressure. Hydrostatic pressure is greatly compounded in the presence of clay soils. Please look again at Figure 4.2. The footing drains around the perimeter of a Log End Cave-type of earth shelter continue along the inside base of the retaining wall, under the backfill. The 4-inch perforated drain tubing should continue on a slightly downward path until it comes out above grade. The daylight end of the drain needs to be covered with a tough rodent-proof screen, such as heavy quarter-inch mesh or diamond-patterned "hardware cloth," something sufficient to keep vermin from inhabiting the footing drain. (See Sidebar for the "soakaway" option.)

LOG END CAVE: BACKFILLING AND RETAINING WALLS

We completed the 4-inch drains to a soakaway (see Sidebar) before starting the *megalithic* wall.

Fig. 9.2: Megalithic wall at Log End Cave.

(*Megalith* comes from the Greek *mega* for great and *lithos* for stone, and *megalithics* is a subject near and dear to my heart, and one that we conduct classes in at Earthwood.) Most of the stones we used are dense sandstones – red, yellow, and purple – that came from the excavation. Not only did we make use of boulders that would otherwise need to be removed from the site, but we wound up with a strong and beautiful wall.

Our backhoe operator, Ed Garrow, had a lifetime of experience with chaining, lifting, and otherwise cajoling large stones to where he wanted them. I made sure I was intimately familiar with my pool of boulders before Ed

arrived, so that I knew exactly which stones would work on the first course, and in special circumstances. I cataloged the stones on a chart, indicating the size of each boulder and whether or not it had good square faces for building, a quality that, thankfully, many of them had. I asked Ed to push the best of the stones to a central location near the work, which saves a lot of running around for a particular stone later on.

Even though our typical stone weighed several hundred pounds, the building procedure was the same as building any dry stone wall. Ed began by clearing a flat base for the wall, depressed a couple of inches to form a bed in which the large stones

Fig. 9.3:
The low
retaining wall at
Earthwood is
about the same
height as the
buttresses
described in
Chapter 3.

could rest. Then, using a heavy chain with hooks at each end, Ed would lift the required stone with his backhoe arm and set it into position. On the first course, we used stones with two parallel faces, saving stones with only one good face for the top course. We shimmed with small flat stones to remove any wobble in the megaliths. These shims can be tightened in place with a hammer. Sometimes, holes between large stones were filled with smaller stones and shims, a detail which shows clearly in Figure 9.2, below. We also used megaliths at Earthwood, although, the wall — typically 36- to 48-inches high — is quite a bit lower than the seven-foot-high wall next to the

door at Log End. Figure 9.3 shows the low retaining wall at Earthwood.

At both Log End and Earthwood, we backfilled with good sandy soil as we built the wall. Thus, the wall was supported by earth from behind. The backhoe was the right tool for both megalithic work and backfilling. Ed would tamp with the bucket as we backfilled the retaining wall.

With any retaining wall, it is good to start with a bit of a ditch or trough, below grade, partially filled with crushed stone. In clay soils, you will want to make sure that this crushed-stone-filled foundation trough slopes out above grade, so that that it doesn't simply fill with water and freeze. With any materials – whether megalithic or one of the choices below – it is good if the materials of the first course extend slightly below grade, helping to "key" the first course into the landscape.

OTHER CHOICES FOR RETAINING WALLS

Drystone Walls

If you are lacking in megaliths, but have a supply of smaller stones – I'll call them *miniliths* – you can still build an effective retaining wall. Drystone wall-building is a fairly skilled task, and only goes quickly at the hands of skilled wall-builders.

Neophytes can certainly do it, however, although you may find it to be a rather time-consuming process. Once again, it may be easier to build the wall at the same time as backfilling, so that the backfill can support the wall as you build. With any retaining wall, it is stronger if the face of the wall angles back towards the earth. It is much easier for the earth to push over a vertical wall, for example, than to push a wall with a 75-degree tilt back towards the berm.

Use your best, flattest, and largest stones at the bottom. Foundation is everything. Keep backfilling, by hand, as you build. Why "by hand"? Well, not many of us can afford to pay for a backhoe operator to stand idle while we do time-consuming drystone wall-building.

Mortared Stone or Block, or Surface-bonded Block

Drystone walls relieve hydrostatic pressure much better than mortared walls. I do not advise the use of mortared stone (or blocks) as a retaining wall, unless you deliberately leave substantial "weep holes" frequently in the wall to relieve hydrostatic pressure. A friend built a surface-bonded block retaining wall at his earth-sheltered home, but, once, during extreme wet conditions, the wall was pushed over by water. If "drainage is the better part of waterproofing" – and it is! – then, equally, drainage is your main ally in preventing retaining-wall failure.

Rammed Tires

First, I have to say that I am not a big fan of interlocking rammed tires as a retaining wall. I appreciate the effort to recycle a waste product, keeping it out of the landfill, but such a wall still looks like … a stack of tires. By ramming the tires with a sledge hammer and with the use of vertical rebar to tie them together, such a wall can be strong and long-lasting. But there are more esthetic choices.

Landscaping Timbers

Landscaping timbers are very popular as retaining walls. You can use pressure treated 6-by-6s, 8-by-8s, or other shapes made for the purpose, such as logs flattened top and bottom. Locust is the only non-treated wood that I am aware of which will not rot out when used against soils. Another

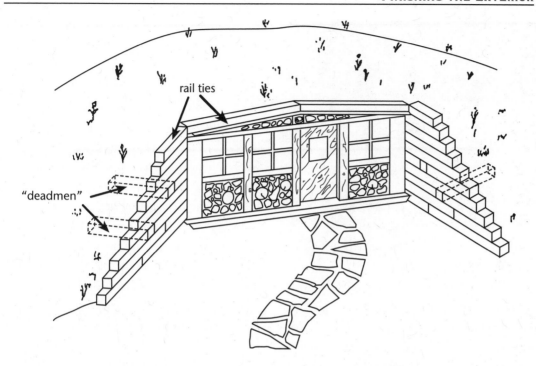

rail ties

"deadmen"

Fig. 9.4:
Timber retaining
walls are tied to
the earth berm
with horizontal
"dead men."

option is recycled railway ties. Some people may not like them any more than tires for appearance, although they are just about acceptable to my esthetic sensibility.

Whatever the type of timber you choose, the technique is the same: Have a good tamped base of coarse sand or crushed stone to start the wall on. This base relieves a certain amount of uplifting frost pressures, and also serves as an opportunity to level the ground for the first course of timbers.

The key to a good timber retaining wall is the frequent use of "deadmen," timbers of the same dimension which run into the earth at right angles to the wall. Figure 9.4 is a schematic drawing

showing the use of deadmen in a retaining wall for a small Log End Cave style earth shelter. Figure 9.5 shows a neighbor's excellent retaining wall of 6-by-6 pressure treated timbers.

COATING EXPOSED RIGID FOAM

As discussed in Chapter 1, we must not have any "energy nosebleeds" conducting heat out of the house. Clearly, this is something that must be attended to way back at the design stage. Two inches of extruded polystyrene, worth R-10, should be applied to the exterior of blocks or concrete work that would otherwise be exposed, effectively preventing the nosebleed. But polystyrene cannot be left exposed to the sun's

Fig. 9.5:
This retaining
wall is made of
6 × 6" pressure-
treated timbers.

harmful ultraviolet rays, or it will turn to powder. It is necessary to protect all exposed rigid foam insulation with a coating of cementitious material. I have successfully used surface-bonding cement at this detail, as well as my old universal mortar-and-plaster standby of three parts screened sand and one part masonry cement. Apply a bonding agent to the rigid foam to get a better bond between the cement and the insulation. I have also, on occasion, tacked one-inch hex-mesh chicken wire to the Styrofoam® to get a better plaster attachment, although the bonding agent works very well.

If only a little insulation is exposed, there is the option of protecting it from UV rays with aluminum flashing.

LOG END CAVE: LANDSCAPING

We could back a car or pickup truck to the front door. After offloading groceries or firewood, we'd park the vehicle in a turnaround area 80 feet from the house. We liked to keep the landscape free of the clutter of vehicles. The approach to the house sloped away from the door and was covered with 6-mil black polythene and 3-inch disks of elm. The space between disks was filled with #2 crushed stone. The disks stayed in good condition for a few years, although they tended to get slippery when wet. The new owners of Log End Cave paved the same area with concrete. Flat stones would have been better than the elm slabs, which were attractive but only a temporary solution.

I surround almost all of my above-grade buildings with a skirt of black plastic covered with a couple of inches of crushed stone. The skirt carries water away from the building and the plastic stops weeds from growing. Paving blocks or flat stones work well for use as pathways.

We had two huge piles of earth to spread out over the Log End site, the larger on the east side. A bulldozer is the best machine for the job. Because we'd taken the retaining wall on the east side all the way to an old stone wall surrounding the meadow, we had an ideal place to spread excess earth: right behind the retaining wall. On the west side, we had the perfect amount of earth to create a natural-looking slope away from the roof. The net effect is that the roofline looks like a natural knoll on the landscape. Lots of people commented favorably on the way that the home melded naturally into the surrounding terrain.

To keep mosquitoes down, we mowed the area around the house all the way to the old stone

Fig. 9.6:
Earthwood, as
seen from
halfway up the
108' windplant
tower.

wall which marked the edge of the meadow. We build a stone circle of boulders surrounding a firepit, just to the east of the home, using the last eight stones left over from the excavation. Years later, we refined our stone circle building technique to include standing stones nearly seven feet high and weighing up to two tons. Outlier stones mark the rising and setting of the sun on important days like the solstice and the equinox … but that's another story and another book.

Beyond the stone wall, on the south side of the home, we left the forest wild, save for the curved driveway which provided delivery access. Also, we removed a few coniferous trees that stopped valuable sunlight from reaching the home in winter. Removing these trees gave us the added benefit of being able to see deeper into the forest, and, occasionally, a deer would wander through our field of view.

EARTHWOOD: LANDSCAPING

At Earthwood, we were really starting with a clear page. Almost two acres of land had been rendered to moonscape by the removal of the gravel layer. Over the years, we have reclaimed almost all of this two acres, and visitors can hardly believe the transformation. We found that just an inch of topsoil, in combination with grass seed and lots of mulch, will return even a gravel pit to living, green production. We built eight raised-bed gardens in front of the home, where Jaki grows a variety of vegetables. The wide treeless expanse in front of the home gives us plenty of solar gain for passive heating as well as for our off-the-grid electric system. And if we get our electrical energy from the sun (and wind), then we get our spiritual energy from the stone circle.

Chapter 10

INTERIOR FINISHING CONSIDERATIONS

Perhaps even more than surface dwellings, earth-sheltered houses require that all or most of what makes a home livable had better be attended to at the design stage. Once the design has been worked out right, and the shell is finished as described in previous chapters, the interior work of an earth-sheltered home is really ...

NOTHING SPECIAL

Plumbing, electric, internal framing, wall coverings, and decorating are really not that much different with earth-sheltered housing. As mentioned in Chapter 9, we did a good part of our framing at the Cave prior to backfilling, to lend extra lateral strength to the walls. Now is the time to do any electrical or plumbing work within the interior wall framework. (See Sidebar, page 197 for electrical considerations that might be a little different with an earth shelter). The temporary bracing can then be replaced with the desired wall covering: wood, sheetrock, tile, or whatever.

The fact that we integrated our floor plan with the structural plan made for easy framing and finishing. Taking Log End Cave and Earthwood together, there was only one short eight-foot wall – at the Cave – that we had to fit around roof rafters, and, as that wall was composed of individual pine boards, it was not difficult to do. In most cases, our internal walls rise up from the floor to meet heavy timbers: girders, floor joists, or rafters. No fitting of wall framing – or wall covering around exposed members – was ever required. In fairness, it must be stated that this is not a problem with homes having false ceilings, either, but if you are using heavy timbers to support the loads involved, there is something comforting about seeing them overhead. They are beautiful. Who is not favorably disposed towards the heavy beams in a good old English pub?

At Earthwood, we were often able to pre-build a wall frame on the floor and tilt it up between posts, as seen in Figure 10.1.

Fig. 10.1:
It is often easier
to frame a wall
section on the
floor and then
stand it up, as
opposed to
toenailing each
stud to the floor
and the girder.

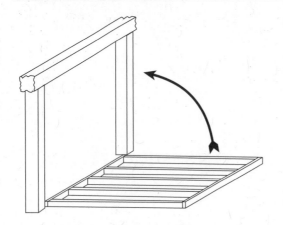

Most earth shelter advocates and builders go out of their way to maximize light in their homes. This is accomplished by judicious planning of windows and skylights (or modern "light wells") at the design stage, and by keeping walls light and bright during interior finishing. Mac Wells maintains that the cheapest energy saving device we have available to us is a gallon of white paint. Recently, this point was driven home to me when our son redecorated his room at Earthwood. There was a dark section of cordwood masonry on the inner surface of an exterior wall, which we decided to cover with an ivory-colored latex paint. The result was striking: the black hole of the room became bright and – yes – beautiful. The texture and relief of the cordwood still came through nicely. We even left a section deliberately unpainted, to show off the effect.

In other rooms of Earthwood, mostly upstairs, Jaki and I were careful to use bright white internal walls of textured paint to reflect light back onto the inner surface of the exterior cordwood walls. An excellent low-cost textured paint can be made by combining premixed drywall joint compound with the cheapest white latex paint, at a proportion of about 4 parts compound to 1 part paint. The paint brightens the mixture and gives it the right consistency for application with a textured roller. This home-made product is only a fraction of the cost of manufactured textured paint, and has looked great at Earthwood for a quarter-century. Downstairs, at the north side of the home, where light is at a premium, we painted our surface-bonded block walls with ordinary white latex paint, and they look good, not at all like blocks.

But, no, we will not be painting all of our cordwood walls white!

FILLING THE SPACE BETWEEN RAFTERS

At Log End Cave, the three long 10-by-10 girders carry the exposed 4-by-8 roof rafters. At Earthwood, 4-by-8 floor joists and 5-by-10 (and a few 6-by-10) rafters are carried by an internal framework of heavy timbers: 8-by-8s, 9-by-9s, and 10-by-10s. In each case, there is a space between rafters or joists that would allow sound to readily transfer from one side of the wall to the other. If we want to minimize sound transfer, we need to fill those gaps in some way.

It was in keeping with our cordwood architecture to fill the space with leftover pieces of heavy timbers, surrounded by a mortar joint. As I write these words, and look overhead where the

Special Electrical Considerations

If you have a concrete floor, the National Electrical Code (NEC) requires that all plugs (called duplex receptacles or DRs) be ground-fault protected, either by use of ground-fault-interrupted (GFI) duplex receptacles or by the use of ground-fault protected circuit breakers back in the electrical supply panel box. With GFI receptacles or circuit breakers the danger of electrical shock in highly conductive areas—such as a wet concrete floor— is eliminated.

In order to meet code, and for practical reasons, it will probably be necessary to have at least a few duplex receptacles (DRs) around the inner surface of the below-grade wall. The NEC says that there must be a DR at least every 12 feet around the perimeter of a room. With a small room, it is sometimes possible to meet code with all of the DRs located in the interior walls.

With concrete, concrete block, or cordwood walls, the easiest way to install DRs is by the use of either metal conduit or code-approved surface-mounted wire mold. Approved wire-mold systems are made by several companies and come in different styles and colors. It is attractive, safe, and easy to add to, repair, change, or fix after installation. It allows the builder to build the outer walls first and wire them after. One manufacturer whose products I've used is the Wiremold Company in West Hartford, Connecticut. Contact them (see Appendix B) for their catalog and wiring guide.

inner wall meets the ceiling of my office, I see the space between the 5-by-10 rafters filled with pieces of 8-by-8 beam, each piece surrounded by mortar. Each chunk of beam is cut to a length and shape such that it can be completely surrounded by a 1-by-4-inch piece of Styrofoam®, nailed to the chunk's interior hidden surfaces, so that the insulation will wind up hidden in the wall. The Styrofoam® also helps hold the piece in place when it is jammed into its space.

Mortar is easy to apply into the one-inch-by-two-inch cavities surrounding the chunk. Simply load the back of a trowel with mortar and, with a pointing knife, push the mortar off the trowel into the cavity. The Styrofoam® in the middle of the wall provides a resistance to the mortar. Use long firm strokes of the pointing knife to quickly smoothen the mortar to a strong, tight, and pleasing appearance. We like to leave the surface of the chunk just a little proud of the mortar

Other alternatives to chunks of beams between rafters are: regular cordwood masonry (much more difficult and time-consuming to fit into these awkward spaces), glass blocks (which can be expensive; look for a bargain), mortared bricks, or drywall sections nailed to a furring strip (picky work!) The advantages of using chunks of beam are: (1) scrap pieces of material are used up in an attractive way, and (2) they are easy to shape for a perfect fit, especially in combination with mortar as described above.

LOG END CAVE: HEAT SINK

I built a solid freestanding stone masonry room divider between the living room and kitchen-dining area. This stone mass was a foot thick, five feet high and eight feet long. In all, the 40 cubic feet of stone weighed about 6,000 pounds, or three tons. Both the kitchen cook stove and the cast iron parlor stove had their backs to the stone mass, so that it was being charged with thousands of BTUs of heat from the stoves' radiant heat. Three tons can store a lot of heat in the winter, helping to keep the space warm even when the fires are not burning. It was like a giant "night storage radiator" so popular in Great Britain. In the summertime, the mass can store "coolth" in the same way. At any time of the year, it was like having a large heat capacitor or thermal flywheel working for us in the very center of the home.

I had never built a freestanding stone wall with mortar before, although I'd labored for a master mason in Scotland and learned a lot by

Fig. 10.2:
The three-ton
room divider at
Log End Cave
helped the
home maintain
a steady
temperature.

background. Use one of the various cordwood mortars described in *Cordwood Building: The State of the Art.*

Alternatively, you can simply use pre-packaged "mortar mix," or make your own mortar from three parts sand and one part masonry cement. With these mixes, however, brush on a coat of Thompson's Water Seal™ (or its equivalent) to all surfaces of wood that will come in contact with the mortar. Do it a day ahead of time. The water seal product will prevent the dry wood from rapidly absorbing the moisture from the mortar, which can cause mortar shrinkage cracks.

osmosis. It took me five-and-a-half days to build the wall and hearth, with Jaki doing most of the mortar pointing using an old butter knife with the last inch of its blade bent up at a 15-degree angle. We even incorporated some useful features into the mass, such as a stone table, shelf, and little bench in the living room, its top surface in the shape of a butterfly.

Our very successful mortar mix was 5 parts sand, 1 part Portland cement, and 1 part masonry cement. Mortar for stonework should be mixed quite a bit stiffer than brick mortar. The stones came from the old stone wall which we'd cut through with the bulldozer to provide vehicular access to the front door.

EARTHWOOD: MASONRY STOVE

Even better than a heat sink charged by radiation from without is a masonry stove (or "Russian fireplace") charged from within. It is briefly described in Chapter 6 and seen in Figure 10.3. See also Bibliography.

FLOOR COVERING OPTIONS

At the Cave, the floor coverings were low in cost and almost purely practical: sheet vinyl in the kitchen and bathroom, and – because of owning two German Shepherds with perpetually muddy paws – industrial-grade low-knap carpeting in the living room. The mudroom and the utility room floors were coated with concrete floor paint.

Fig. 10.3:
The masonry
stove at
Earthwood.

Hardwood

You can put a wooden floor over a concrete floor by a variety of methods, but this choice somewhat diminishes the value of the concrete floor as thermal mass, as does carpet with foam under-layment. Hardwood flooring can be applied directly to the concrete floor, in square parquet-style tiles or in short interlocking planks. These systems generally require the use of mastic made for the purpose.

Fig. 10.4: Jaki points around slates set into the fresh concrete at Stoneview, while intern Nick Brown hand-trowels the triangular spaces between the eight slate "spokes."

Concrete Stains

There have been great improvements to concrete floor finishing in the past ten years. A variety of concrete stains are readily available on the market, including etching and penetrating stains, deep glosses, and others. Etching stains, although beautiful, must be professionally applied. These are very hazardous products and, as of 2005, are banned by regulation in New York and several other northeast states. Surface preparation and final finish depends on the product you select and your application. Some products require etching the floor first with a weak muriatic acid solution. Consult with your local paint supplier for color charts as well as characteristics and requirements for the various products.

The Recycled Slate Floor

This floor option deserves its own heading. For over 20 years, I have been enthusiastic about a beautiful hardwearing floor made of recycled roofing slates set in either freshly poured concrete or, as at the Earthwood house, in a ⅜-inch bed of mortar. With small buildings, such as our guesthouses, the technique is as follows:

1. On the same day you pour the slab, prepare the ¼-inch-thick recycled roofing

slates by applying a bonding agent such as Acryl-60® (Thoro Corporation) or an equivalent to the side of the slate formerly exposed to the elements. Allow the bonding agent to dry. The long-protected underside of the slate will become the floor's top surface.

2. Pour the slab.

3. Stretch nylon lines across the pour, from forming board to forming board, to indicate the lines where you want to install slates.

4. Using a rubber mallet, tap the slates gently into the fresh plastic concrete.

5. Wait a while (an hour or more) for the concrete to achieve a consistency that can be pointed with a knife. Pointing too early leads to a soft, weak coating of wet concrete slurry (sand, water, and Portland cement). If the concrete reaches an advanced set, pointing becomes difficult.

We have used a variety of patterns in the concrete, with various percentages of slate. La Casita's floor is entirely slate, with one-inch pointed joints between. The Straw Bale Guest house has a rectangle of 49 slates in the middle, with about two feet of hand-troweled concrete all around the pattern. At Stoneview – see Figure 10.4 – eight spokes of slate all meet together at the building's center post.

In 2006, after this book has gone to press, we'll seal the slates with slate and tile floor sealer,

and select a user-friendly stain for the large triangular areas between the slate spokes. It should look great.

The round 1,000 square-foot concrete slab floor at Earthwood was far too big to attempt slate installation while the concrete was still plastic. Instead, after living with a concrete floor for two years, we installed the shaped slates in a beautiful circular pattern around the masonry stove (although we began at the outer edge and worked towards the middle.) Again, we prepared the slates as per (1) above, but then proceeded as follows:

2. We applied bonding agent to the smooth power-troweled concrete floor.

3. Using a wooden lath as a screeding guide, we laid down about a ⅜-inch bed of fairly strong but slightly wet mortar. The mix was 5 parts sand, 1 part Portland cement, 1 part masonry cement.

4. With a rubber mallet, we tapped the slates gently into the fresh mortar.

5. Excess mortar rising between the slates was scraped away with a trowel and, after waiting a while for the mortar to set, the 1-inch joints between slates were pointed. Do not point until the mortar is quite stiff.

The roughly 650 square feet of slated area took Jaki, son Rohan and the author five-and-a-half days to complete, and we were at each other's

*Fig. 10.5:
The slate floor
draws one's
eye to the
masonry stove
at Earthwood's
center.*

throats by the end of the job. But it is a beautiful floor. We initially applied two coats of sealer to the floor, and renew it with a fresh coat every two or three years.

Recycled slates are not always easy to find, although I have never had too much trouble in northern New York, where they are frequently torn off of old buildings. An alternative would be to watch out for a deal on 12-inch-square floor tiles. Bruce Kilgore recently tiled his concrete floor at his cabin for less than a dollar per square foot by keeping his eyes and ears open for a bargain.

Keep in mind that concrete is hard on the skeletal system, as an hour's shopping at the mall will quickly confirm. Slate, vinyl — even hardwood – doesn't help much. Consider the use of rush mats, throw rugs, or padded carpet where you'll be doing a lot of standing. Or make the decision early on to do a wooden floor, as described in Chapter 4.

EARTHWOOD: THE UPSTAIRS FLOOR

The downstairs ceiling is the upstairs floor: 2-by-6 tongue-and-groove spruce planking. Downstairs, we see an attractive V-joint overhead. Upstairs, the boards butt together. To finish the floor, which was discolored by long exposure and work traffic, I borrowed a floor sander and sanded for about six days, although it seemed much longer at the time. We decided to go with an oil-based floor finish that penetrates into the wood, while also helping to harden it. Rigid coatings like varnish and urethane tend to chip away on a softwood floor, and, at the least, require constant maintenance. Once, in 24 years, we have resurfaced the floor, with a floor sander and a couple of new coats of hand-rubbed oil. If a wear pattern develops at high traffic areas, it is easy to feather in additional oil without having to redo the entire floor.

EARTHWOOD: KITCHEN CABINETS

The curved walls at Earthwood do not present any special decorating problems, because cordwood masonry comes already decorated and the curved surface-bonded block walls look great with a single coat of paint. And interior walls are all straight, so any surface can be used. The only time the curved wall slowed us down was when we built the kitchen cabinets, which we fitted to the curved external wall. Scribing and custom-fitting may have cost us an additional 25 percent in hired labor and materials, but the total cost of our cabinets was only $650 in 1982. And they are quite expansive: eleven large cupboards, a four-drawer unit (bought new for $109), 56 square feet of laminate countertop, a stainless-steel sink unit, and display shelves.

Jaki and I saved a lot of money by making our own cabinet doors and surface-mounting them to the framework crafted by our friend, Bob Smaldone. (Yes, a lot of friends helped at Earthwood. Whenever you can enlist help from someone who can do a job twice as well and in half the time as you can yourself, jump at the chance. It is an even better deal if you can barter time and keep a lot of paper-based busybodies out of the equation.)

The doors were elegantly simple to make and to hang. We bought an attractive "finished" sheet of the best ¾-inch plywood, pine-finished both sides to match Bob's one-by-six pine cabinet surface. With a table saw, we accurately cut the plywood into twelve doors, each measuring 16 by 24 inches. This took only ten minutes, as there were minimal cuts and absolutely no wastage. All we had to do was to sand and oil the doors. So each of the doors cost about $4. Even doubling the numbers for 2005 prices, that's still cheap for cabinet doors. By surface-mounting them instead of flush-mounting, fancy fitting was avoided and labor costs lessened. The attractive results can be seen in Figure 10.6.

By the way, a 16-sided building, instead of truly round, would have made the cabinets go a

lot easier, particularly the fitting of the laminate countertop.

Fig. 10.6: Kitchen cabinets at Earthwood.

CLOSING COMMENTS

Furnishing and decorating are matters of individual taste, and I've no sage advice to offer except that it is a good idea to keep the floor, walls, and furnishings as light and bright as possible. An exception would be a floor intended for direct solar gain. In this case, slate would be perfect, as its darker surface will absorb heat much better than, say, light-colored floor tiles.

Finally, note that a lot of the commentary on finishing in this chapter makes use of rustic detailing: slate and oiled wooden floors, exposed timbers, cordwood and stone masonry. Much of the so-called "finish" work is completed during construction, and the result is a warm and cozy atmosphere. I tell cordwood students that the key to satisfaction with the build quality and the completed home is in the terminology you use. Use words like "rustic," "textured," "character," and "relief" and you will be met, as Thoreau says, "with a satisfaction unexpected in common hours."

Chapter 11

PERFORMANCE

Part One: Log End Cave

We lived at the Cave for three years. During that time, outside temperatures varied from -40° F to 92° F (-40° to 33° C), while the interior temperature ranged from 58° F to 77° F (14.5° to 25° C). The drainage, waterproofing and structural systems met a severe test one spring when it rained steadily for three days on top of 36 inches of already compressed snow. The only problem during that time was one or two minor flashing leaks at the base of the large front windows, easily repaired. And, as already discussed, I no longer advise berming up to the underside of windows on the south side.

I will list other problems that occurred, followed by their corrections or, better, their avoidance:

ENERGY NOSEBLEED – ONE
The block walls at Log End Cave showed to daylight on both the east and west side of the home, classic energy nosebleeds. We didn't notice a problem until December of our first year in the home. Our master bedroom wall, in the southwest corner of the home, and the mudroom wall (southeast corner) were getting wet at their southern ends. Our first thought was that the tricky waterproofing details at these corners were leaking. Fortunately, before tearing into the December earth on a fruitless search for a non-existent leak, I remembered something I'd read by Mac Wells about conduction of heat through parapet walls and the need for a thermal break. See Figure 11.1.

I covered the exposed exterior of the block wall with 2 inches of polystyrene insulation. Within two days, the wall dried up, its temperature now able to rise above dew point. An insulation detail should have been included at the design stage, such as the installation of extruded polystyrene during construction, held in place by the retaining wall and/or mechanical fasteners. Then – and this is imperative – the insulation covered with a protective coating to prevent UV damage.

Fig. 11.1:

Rigid foam on

the exterior of

the exposed

block wall

slowed heat loss

to the exterior

and kept the

internal wall

temperature

above dew

point.

ENERGY NOSEBLEED – TWO

In May of our first year in the home, we began to notice condensation around the base of the block walls. Because of the so-called "thermal flywheel" effect, the earth's temperature in May and June is still quite cool at six feet of depth. This "coolth" is conducted through the footing to the area where the poured floor meets the block wall. Warm moist interior air hits the wall, followed by: dew point, condensation, damp. The condition was the result of our failure to insulate right around the footings and under the floor.

We didn't have a lot of choices regarding a cure. We rolled the carpet back so that the rug wouldn't get mildewed. We used some moisture-absorbent material in an effort to dry things out. Had we been connected to commercial electric lines, we would have used a dehumidifier. We toughed it out, keeping in mind architect John Barnard's warning that earth-sheltered houses experience high humidity until all the concrete cures. "Don't get nervous," he said. "Whenever possible, leave the building open on dry days. Drying out may take two years."

Compounding our high humidity was the fact that our 4-by-8 hemlock rafters and our ceiling planks had been trees just days before being milled, and we installed them shortly thereafter. With our hemlock, which weighs 110 percent more green than dry, we might have had a couple of tons of water trapped on the living side of the waterproofing membrane. The moisture can only transpire into the interior ambient air, not directly to the exterior. I knew that wood takes about a year per inch to dry through side grain. Using the woodstoves – a dry heat – helped a lot, and we ran them every couple of days that first June.

Humidity was lower during the second spring, but there was still some condensation. The third year was even better, but we still had some condensation in the corner rooms. At Earthwood, we discovered that an ounce of prevention is, indeed, worth a pound of cure. We wrapped the entire footings and floor with extruded polystyrene, as described in Chapters 3 and 4, and we have not had any condensation. Humidity levels are comfortable, upstairs and down. The fact that 60 percent of Earthwood's external walls are cordwood masonry helps a lot in this regard, as cordwood homes are notoriously dry.

Leaks in Skylights and Chimneys

Skylights often leak on a conventional roof. With the earth roof, the chances of leaking are even greater. The reader may recall that we waterproofed the original Cave roof with a difficult layering of black plastic roofing cement and 6-mil black polythene. We had leaks around skylights that had to be fixed, but we could have avoided the leaks if (1) we had used a better waterproofing membrane in combination with the flashing provided with the skylight and (2) we had used good positive drainage.

There are lots of layers in the earth roof – earth, insulation, drainage layer, etc. – and it is necessary to box out for the skylight in order to take it above grade. Figure 11.2 shows an appropriate wooden box for the purpose. The Bituthene® or equal waterproofing membrane goes up the sides of the box and the skylight's own flashing comes down over the membrane.

But did you know that "Drainage is the better part of waterproofing?" Figure 11.3 shows the Earthwood chimney surrounded by crushed stone, which meets with the drainage layer beneath the soil.

"Solar tubes" (sometimes called light wells) are highly reflective roof-mounted light tubes that have become popular in the last few years. They come in various diameters, such as 10-inch, 14-inch, and 24-inch. My neighbor installed one over his formerly dark dining room area, and the room's daytime transformation from black hole to bright space is amazing. Two major manu-

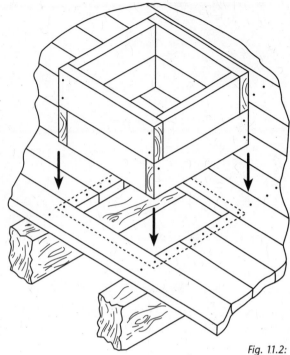

Fig. 11.2:
A wooden box-frame, as shown, is needed for the installation of skylights or light wells. The depth of box must accommodate all of the various layers of the earth roof.

facturers (Solartube and Velux America), are listed in Appendix B, but an internet search ("solar tubes") will list other manufacturers and wholesale sources for purchasing at discount.

The same installations for skylights, above, are applicable to solar tubes.

Pressure on Retaining Board

We retained the soil on the north and south "gable" ends at the Cave with 2-by-12 retaining boards, waterproofed and flashed over their tops. Hydrostatic and frost pressure exerted a force on these retaining boards, so that by the first springtime, the 12-inch-high boards were an inch

Fig. 11.3: Any projection through the earth roof – skylight, chimney, vent stack – should be surrounded by good drainage material, such as #2 crushed stone.

out of plumb. Once again, bad drainage was the culprit. We solved the problem by carefully removing the topsoil next to the flashed retaining boards and replacing it with crushed stone, drained to the edge of the building.

Even though you will see detailing in the literature for retaining boards at the roof edge, my strong feeling now is to leave them out in favor of either sods around the edge or treated retaining timbers supported by 1-inch pressure-treated shims. Both of these methods are described in Chapter 8. The exception would be steeper pitched roofs, such as Chris Dancey's, also in Chapter 8.

LIVABILITY

The quality of life at Log End Cave – its livability – exceeded our hopes. The house was infinitely lighter and brighter than Log End Cottage, built on the surface. Heating and cooling characteristics

were phenomenal. The view to the south was excellent. There were unexpected joys, too, like the sun's rays providing natural illumination to the dartboard on summer evenings, and being able to lie back in the bathtub, bathed by sunshine from the skylight above.

Our goals had been to create a home with superior heating and cooling characteristics; to own it ourselves instead of the bank owning it for us; and to maintain the beauty and natural harmony of the land. All of these goals were realized. Details follow.

HEATING AND COOLING

Over a three-year period, our average fuel consumption at the Cave was between 3 and 3¼ full cords of medium-grade hardwood, equivalent to 360 gallons of fuel oil or about 9,500 kilowatt-hours of electric heat. We estimate that 95 percent of the wood was burned through our efficient Oval cookstove from Elmira Stove Works. We would only occasionally fire the upright Trolla parlor stove, mostly to charge the thermal mass of the north wall during a cold snap, or simply to enjoy a living fire with the stove's front door open.

We noticed that the Cave required less fuel when there was a good blanket of snow on the roof. The winter of 1979–80 was considerably warmer than the record-breaking cold winters of 1977–78 and 1978–79, yet we used more wood because of the total lack of snow until mid-February.

We went away for a month's vacation in 1980, departing February 11th. There was no heat in the

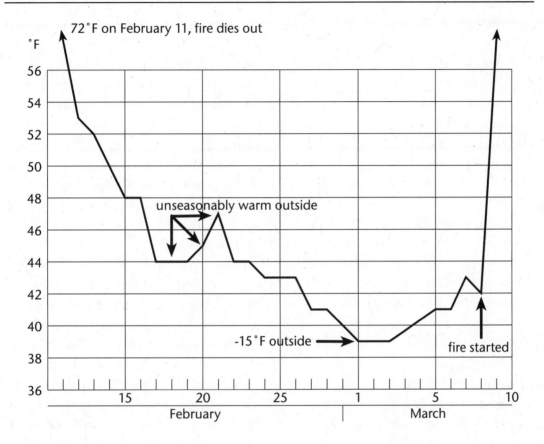

72°F on February 11, fire dies out

°F

Fig. 11.4:
A winter with-
out heat at Log
End Cave.

Cave except for solar gain and the BTUs produced by two large German Shepherds. The windows and skylights were not shuttered. The house reached its lowest temperature of 39° F (4° C) on March 1st (with the outside temperature at -15° F [-26° C]). After that, the house temperature began to rise. See Figure 11.4. The temperature was always taken at 7:30 a.m., before any solar gain, the coldest time of the day in the house. On sunny days, the internal temperature would rise 6° F to 8° F (4° C to 5° C) by 2 p.m.

Figure 11.4 illustrates that earth-sheltered homes, set into above-freezing earth temperatures, can be left for extended periods without the need for back-up heating to protect pipes from freezing.

Living at Log End Cave also confirmed three points made in Chapter 1 about summertime cooling:

1. The home benefited from the evaporative cooling effect of the earth roof.

2. The earth's coolest temperatures lag almost two months behind the climate's lowest average temperatures at a depth of 5.3 feet (1.6 meters). Even at the height of summer, surrounding earth temperatures are cool.

3. The earth-sheltered house enjoys a slow temperature change, due to the "thermal capacitor" effect of its own great mass in combination with the even greater mass of the surrounding earth. This moderating effect effectively averages out temperature variations over a day, a week, or even a month. The Cave showed that temperature curves for an earth-sheltered home have gentler slopes than those for surface dwellings. Earth-sheltered homes, at least in the North, do not require air-conditioning, although they may require some form of dehumidification, particularly during hot humid periods.

ECONOMY OF CONSTRUCTION

The building methods described in this book are sound methods, they are readily learned by owner-builders, and they are moderate in cost. Please refer to Table 4.

Because we kept receipts for virtually everything, I'm certain that the larger figures at the bottom of the table are close to reality. The totals do not include the septic system, well, windplant, storage batteries, or furniture, all of which were already in use at the Cottage.

Nor do the figures place a dollar value on the owner-builder's labor, but I maintain that this value should be measured in terms of time, not money. Roy's First Law of Empiric Economics is this: "A dollar saved is much more valuable than a dollar earned, because we have to save so darned many of them to save so precious few."

In 2005 dollars, you could figure $22,000 to $33,000 (about $22 to $33 per square foot) to accomplish the same thing, depending on where you live and who you are. (See also Cost at Earthwood, below.)

About 1,800 man (and woman) hours were required to build Log End Cave, including 450 hours of outside help, some of it paid and some volunteered. This labor estimate may not be quite as accurate as the cost estimate, although I feel it is within 200 hours of reality. Each builder works at a different rate anyway. Most of the work was done during a five-month period ending on December 17, 1977. We completed most of the interior work during the first two months of 1978. Final landscaping was finished in June.

Part Two: Earthwood

Back in Chapter 1, I enthused: "My father used to say you had to build two houses to get one right, one to make all your mistakes in. Well, my father was cleverer than me. It has taken us three houses to get one just the way we like it, and Earthwood is that house."

Table 4
Log End Cave Cost Analysis

SERVICE OR PRODUCT	COST (US$)	SERVICE OR PRODUCT	COST (US$)
Heavy-equipment contracting	$892.00	Water pipe	$61.59
Concrete	873.68	Metalbestos® stovepipe	184.45
Surface-bonding cement	349.32	Rigid foam insulation	254.93
Concrete blocks	514.26	Roofing cement	293.83
Cement	47.79	6-mil black polyethylene	64.20
Hemlock	345.00	Flashing	27.56
Milling and planing	240.26	Skylights	361.13
Barn beams	123.00	Double-pane insulated windows	322.50
Other wood	167.56	Interior doors and hardware	162.80
Gypsum board	72.00	Tools, tool repair, tool rental	159.43
Particle board	182.20	Miscellaneous	210.31
Nails	62.88	Materials and contracting cost	
Sand and crushed stone	148.21	of house, subtotal	$6,750.57
Topsoil	295.00	Paid labor	660.00
Hay, grass-seed, fertilizer	43.50	Cost of basic house	$7,410.57
Plumbing Parts	124.95	Floor covering (carpets, vinyl, etc.)	309.89
Various drainpipes	166.23	Fixtures and appliances	507.00
		Total spending at Log End Cave	$8,227.46

Does that mean the house is perfect? No, but there only a few very minor changes I would make:

1. I wouldn't bother with earth tubes if I were building a similar house in the North again. And this is why I didn't describe them in this book, even though they did appear in my *Complete Book of Cordwood Masonry Housebuilding*. They are simply unnecessary. The house stays nice and cool in the summer without them. But the masonry stove and the other wood stoves do need air to operate

properly. So, instead of earth tubes, I would bring a direct 6-inch-diameter pipe right to the side of the masonry stove firebox, as described in Chapter 4, in the part called Under-Stove Vents.

2. I might go with a lighter-weight living roof, as we have done at Stoneview, and described in Chapter 8. A less heavy roof would allow the use of smaller, lighter, less expensive rafters ... or greater spans with the same size of rafter ... or some combination thereof. A lighter weight earth roof can be accomplished by the use of composite drainage materials and/or less soil. Using less soil requires a careful selection of plants, such as sedum and chives, which will take the living roof through a drought situation.

3. Jaki and I will probably never build another Earthwood home again, but if we did, we would build it within the confines of a 16-sided timber frame, as described in Chapter 6 of *Cordwood Building: The State of the Art*. The advantage of working under the protection of a roof in place cannot be over-emphasized. And the timber framing aspect may very well ease your passage through the permitting process. Unlike an octagon, a 16-sided building looks and feels like a truly round building.

COST AT EARTHWOOD

I did not keep as accurate an accounting as I did at Log End Cave, but I feel confident that I am within 10 percent of the reality with these figures. The total *value* of the labor and materials for the original Earthwood house, including the downstairs solar room, was right around $20,000, or about $10 per square foot in 1981–82. This estimate includes our simple plumbing and electric internal house systems, but not the well, septic system, photovoltaic cells, or windplant. — just the house.

In reality, we spent a lot *less* than $20,000, as we received major donations of Dow Styrofoam®, Bituthene™ membrane, and Conproco® surface-bonding cement, which, all told, were probably worth $6,000. In exchange for these materials, we tested them in our earth-sheltered home and reported on their use. I've learned this in life: It never hurts to ask. I speak well of all of these products because they have performed well for us, but I am under no obligation to endorse them.

In 2002, we added a 200 square-foot sunroom to the building, over the existing solar room. Its construction is detailed in Chapter 5 of *Timber Framing for the Rest of Us*. The room, which we love, includes five expensive windows and a new pre-hung interior door. Nevertheless, we came in at about $4,000 for materials and $250 for hired labor, about $22 per square foot total out-of-pocket cost. We did not enjoy economy of scale on the sunroom, but, on the other hand, we built over an existing room, saving on foundation cost.

Now Hear This!

More than a third of the average American's after-tax income is devoted to shelter, usually rent or mortgage payments. If a person works from age 20 to age 65, it can be fairly argued that he or she has put in 15 years (20 in California!) just to keep a roof over their head. With a piece of land, six months' work, and – say – $35,000, he (or she) and his family could have built his own home.

To save 14½ years of work, you cannot afford *not* to build, *even if it means losing a job while you do it.* Granted, the land (and the $35,000) has to come from somewhere, but this amount is no more (and probably less) than the down payment on a mortgaged contractor-built home, and about half the cost of a new double-wide mobile home (figuring either option as being about the same square footage as an earth-sheltered home).

And what do you get for your time and money? You get a comfortable, long-lasting, energy-efficient, environmentally compatible, low-maintenance home. You get the design features that suit you, so that the house fits like an old slipper. You get built-in fire, earthquake, and tornado insurance. You get intimate knowledge of the home so that when maintenance or repairs are required, you're the one best placed to make them. You get tremendous personal satisfaction. And you get freedom from a lifetime of economic servitude.

I believe that I could build Earthwood today for about $20 per square foot. I'm a pretty good scrounger and make good use of indigenous, recycled, and bargain materials. And we are not over-regulated in our little town in northern New York. I think other owner-builders could do a similar home for $20 to $40 per square foot, materials cost. The wide parameters in that figure are because: (1) prices vary greatly around the country and (2) everybody is different, with different desires and levels of acceptability.

The sidebar above is an almost verbatim reiteration of something I wrote for the long out-of-print *Underground Houses: How to Build a Low-Cost Home*. Now, a quarter century later, I still feel the same way.

But be careful out there. And, I refer not to safety on the job site, which, of course is imperative. Rather I am talking about working within your capacity ... or capacities, as what I am about to tell you is even more important with couples than with individuals, as there is a relationship at risk. Listen:

ON BIG PROJECTS

For several years before building Earthwood, I'd been aware of the danger of tackling big projects. I'd seen several marriages and relationships go sour during the seemingly never-ending construction process. Lots of people – maybe most – have difficulty with living in a tent or cramped quarters for too long, or, equally stressful, actually living in a construction zone. But, with two houses already behind me, I thought I was proof against any such concerns in my relationship with Jaki.

But Jaki knew better. She was not surprised when the pressures of housebuilding began to take their toll. But neither of us was prepared for the debacle of having to rebuild cordwood walls – twice – because of dry hardwood expansion, a situation we'd never encountered before. The old familiar scenarios returned: rushing to beat the winter, financial pressures, and just plain disenchantment with the process. Two things got us through the project with an intact marriage: Jaki's patience, and the fact that we had the comfortable Log End Cave to return to each evening.

We moved into Earthwood 13 months after breaking ground. The upstairs portion, a complete apartment on its own, was comfortable, although we drew water from the outdoor pitcher pump for five months before completing our bicycle pump system, which we still use 25 years later. Completion of the downstairs took longer— another 18 months, in fact. But the pressure was off and the pace slowed naturally.

I feel that a project on the scale of Earthwood (2,400 square feet gross using alternative building methods) is too much for inexperienced owner-builders to tackle. No, the techniques aren't that difficult. The house is simply too big. A one-story version would be a much safer bet. The home could be capped and, later, perhaps, a second story could be added. This might be the more affordable strategy, too, if freedom from mortgage is important to you. (It should be, if you don't mind my saying so.)

But, in truth, most people who build their own home get a high degree of personal satisfaction from their accomplishment. So I would like to conclude this book on a less somber note, the true case study of former students Mark Powers and Mary Hotchkiss, who found their earth-sheltered house project to be a very spiritual event. In Mark's words, they have created a sanctuary. Turn the page and read all about it.

And I'll see you later … in Appendix A.

Our Earth-Sheltered Home – A Case Study

by Mark Powers

Since my wife Mary and I moved to Michigan's North Country ten years ago, the idea of building our own home has been the focal point of our lives. In August of 2005 we completed the final stages of a long process of research, building, and learning...and settled into our new home.

To me, though, the home is much more than a house. I describe it as "a cross between a work of art and a force of nature." Here is our story:

We started with researching alternative building methods, making many a trip to bookstores and the local library. Geodesic dome designer Buckminster Fuller was a significant early influence, so my first winter was spent meticulously constructing scale models of domes complete with miniature pieces of furniture and cutouts of Mary, the cats, and myself. Though time-consuming, building a scale model of a house gives a feel for space, and we did the same with many subsequent home designs.

John Larch, a co-worker, helped us to focus on our dream. Over some years, John had used one of Rob Roy's books to build a two-story cordwood home high on a hill. To walk into his house was to fall in love with the coziness and earthy delight of cordwood masonry. A three-day workshop with Rob and Jaki cinched the deal for us. We could do this!

At that point we had only seen pictures of living roofs, but we knew we had to include one in our design. There's something appealing about having grass overhead and never having to shovel the snow off. Building an earth-bermed house also appeared to be a natural way to save energy and blend our home into its hillside setting.

Our 40-acre parcel is heavily wooded and extremely hilly, not a level space to be found. It is also three-quarters of a mile off the main road and accessible only by an abandoned power-line easement. Our initial task was to turn the first half-mile of easement into a passable driveway and level a spot on a ridge for a mobile home site.

We purchased an old mobile home for $1,500, allowing us to move to the land quickly and with a minimum of investment. Moreover, this "temporary shelter" freed us from paying rent. We were able to take our time familiarizing ourselves with the land. And, as a former English Lit major, I needed time to develop practical skills such as becoming proficient with my chainsaw and chainsaw-powered lumber mill. Many of the mature trees lining the old power line easement were turned into timbers for the house-to-be, and stored aside over the course of our first three years here.

We spent six years in our mobile home. With a woodstove and lots of weatherproofing, it was a cozy shelter, even in the face of northern Michigan winters.

We selected our site and designed the house over the course of these first three road-building years. Many a winter's day was spent tromping various house footprints in the snow. In better weather we put up strings and stakes in order to sketch out variations on our plan. Pencil and paper came later. There's nothing like an actual-sized house plan to help get those creative juices flowing. Eventually, we settled on a one-story home based on Rob's 40-by-40-foot Log End Cave design.

From the outset I made a point of introducing myself to the supervisor of our building department. We developed a cordial relationship, one that paid dividends later when our intended house design failed to fit conveniently into the confines of our county's relatively stringent building code. I made it a policy to approach all

the building inspectors as potential allies, and they in turn responded on numerous occasions with helpful feedback. As an "amateur" builder, I had a good deal to gain from their considerable experience. Oftentimes we found ourselves in uncharted waters relative to established code, but because of my positive attitude the code officials were willing to work with me and consider unconventional building strategies in a positive light.

All bureaucracies are inherently conservative, however, so I, in turn, had to do some stretching. A cordwood wall, from their point of view, didn't constitute a quantifiable, load-bearing element, but a timber frame with cordwood masonry infill was acceptable. All the hardwood timbers I had so laboriously milled, although virtually defect free, had to pass inspection by a certified timber grader. It took several months to find someone who had the know-how and credentials to grade hardwood timbers for structural applications.

The building department's concern about loading our foundation and timber frame with an earth roof brought yet another professional into the mix. It was required that our plans pass muster with a structural engineer. Engineers, too, are a relatively conservative group, probably a good thing when you think about it. After one false start, we found a structural engineer with some experience in timber framing and a willingness to consider our project with an open mind. He signed off on our foundation and framing plan and helped us reassure the building department's supervisor as to the soundness of our design. The engineer's services added roughly a thousand dollars to our project – worth it in the long run as he was always there with a quick response as the work proceeded and additional concerns and changes came up.

With absolutely no building experience, I started small, first with an 8-by-12-foot addition to our mobile home, then, an 8-by-16-foot shed. Next, we built a 24-by-32-foot garage, timber-framed with cordwood infill, and a metal roof. This building allowed us to gain experience and confidence in the techniques we'd utilize on our home. It also allowed me to demonstrate some of the basic components of our house design to an interested yet skeptical group of building inspectors. For them, the stakes were much lower when it came to building a garage.

At this point we had the good fortune to strike up a friendship with Les Arnold, a licensed builder and a competent electrician and plumber. He had "been there, done that" as far as building an earth-friendly home goes and has become my guru in most things building-related. For the amateur independent builder, a friendly source in the trades is an invaluable relationship to have. While Rob has served as a generous consultant along the way, there's no substitute for on-site expertise. Les had stepped in on a number of occasions and saved us a lot of time and potential headaches.

Our experience as independent, alternative-minded builders has brought about a number of unanticipated benefits, the most rewarding of which was the "2002 Powers Family Cordwood Reunion." Nearly everyone in my large extended family showed up that summer for a solid week of hard labor. I was amazed and touched by their enthusiasm for the task at hand. By the end of our "workshop" together we had successfully and professionally polished off all of the cordwood infill between the posts and girders on our garage. Everyone went away tired, but pleased with their accomplishment. The nearly unanimous verdict was: "Best family reunion ever!" I felt like Tom Sawyer. Two summers later, the same crew was back at it, laying up the cordwood on a much more ambitious project: our house. I would wish this kind of family on everyone.

I'll cover some of the building strategies we employed, particularly those related to earth sheltering.

The south-facing ridge where we set our home overlooks a small valley with a year-round spring. The north wall of the house is buried to within two feet of the roof eaves at the corners. The east and west walls step down quickly after the first ten feet in order to allow for windows and a main entrance door roughly centered on the east wall. We wanted the earth's protective qualities, but wanted to avoid the cave-like effect that could result from completely burying the house on three sides.

We invested in a thousand dollars worth of radiant in-floor plastic tubing that we included in the slab pour, and installed an appropriate-quality

Mark Powers

*Fig. 12.1:
Mark employed
a foam-block
forming system
for his
reinforced
concrete walls.*

dual-purpose water heater. With radiant floor heat and a centrally located wood stove, we feel we have the best of both worlds.

We selected a foam-block, poured-concrete wall system for our below-grade walls. (Amvic Insulating Concrete Forms, see Appendix B). Although more costly than a conventional concrete-block wall system, it poses several advantages, not the least of which is a considerable savings in time. We set up the forms in a single morning and, with a pump truck, poured the walls in just one-and-a-half hours. We opted for an 8-inch-thick concrete wall for stability and strength. With 8 inches of poured concrete we eliminated the need for any additional reinforcement along the 40-foot north wall, such as pilasters or internal "shear walls." The foam blocks provide us with the desired thermal break between the wall and the earth. On

the inside, we left the foam in place. The foam-block system we employed has regularly spaced plastic webbing that acts as a support structure for hanging drywall, and the walls involved (inside the utility room, bathroom, and part of the master bedroom) are isolated from any direct input of radiant energy from the woodstove.

Another major advantage to the foam-block wall system is that it saves a lot of wear and tear on the body. (Having back surgery makes one especially sensitive to this particular advantage.) An additional – and unexpected – advantage of going with the foam-block system was that the same company that sold us the block was able to order 4-by-8-foot sheets of rigid foam in any thickness up to 48 inches. We settled on a five-inch sheet for our living roof, at a considerable savings in cost and installation time compared with overlapping several layers of 1- or 2-inch Dow Styrofoam® Blueboard™.

As mentioned, we went with heavy timber framing on above-grade walls to satisfy code officials. Our cordwood walls are 16 inches thick, but the posts are 8 inches square. We centered the posts in the walls, which left a four-inch cavity in the exterior face of the wall that we flushed out with 2½ inches of rigid foam overlaid with an exposed vertical 8- by 1½-inch-wide plank. This "faux post" was well-anchored to the real one with heavy screws, counter-sunk and plugged. This provides the walls with additional insulating value and maintains the pleasing post-and-beam façade on the exterior. On interior walls, per Rob's suggestion, we flared back all log-ends adjoining posts with a 45-degree angle cut back to the post, an attractive detail.

The foam block system led to one innovation of note. Our concrete wall, with its foam insulation, was 13 inches thick. That meant that our 16-inch log-ends, centered on the concrete wall, would stick out 1½ inches on both the inside and the outside. In order to create a 16-inch-wide base to carry the cordwood wall, we laid 2-by-6-inch cedar planks on their wide edges, running parallel with the wall, and with 1½ inches of the planks overhanging on each side. See Figure 12.2. These two planks laid on top of the concrete wall system were tied together at roughly two-foot intervals with little 2-by-4-by-6 cedar blocks, toe-screwed to the planks. The rest of the four-inch space between planks was filled with insulation. The cedar planks widen and stabilize the wall, and create a nice trim.

We made one departure from the 40-foot-square framing plan as it appears in Chapter 1 of this book: we eliminated a post from the middle of our living area. This change created a clear span of a little over 18 feet at the ridge and necessitated a call to our structural engineer. He recommended a girder, made up of two engineered paralam beams, each 5½ inches by 16 inches. The two beams are laid side by side and thru-bolted every two feet, creating an 11- by-16-inch composite girder. This massive beam dominates the center of our living space, and, with several coats of linseed oil, has its own unique and

Mark Powers

Fig. 12.2: The poured wall is broadened to accommodate the 16-inch cordwood wall by the use of this 2 × 6" detail, described in the text.

Mark Powers

Fig. 12.3: Mark and Mary in their new home. The engineered paralam beam overhead spans 18 feet.

Fig. 12.4: Mary and the ladies spread earth upon the roof. Mark's Kubota tractor gets it close.

intriguing character. It cost about $900. See Figure 12.3, below.

We decided against completely enclosing our spare bedroom because it has a large window that adds needed light into the main living area. Instead, we built up the walls to a height of 40 inches using 8-inch cordwood to match the post size. We installed additional 8-by-8 posts to frame

the entryway to this room. We'll install bamboo curtains or shutters of some kind to create a privacy option for guests.

A tractor – a four-wheel drive, 45-horsepower Kubota – has become this man's best friend. Early on we rationalized the considerable cost of a new Kubota (with backhoe) in terms of the money it would save us in all phases of the building process. I was able to do my own footing trenches as well as power and water lines. The Kubota allowed me to single-handedly erect the timber frame. (A 12-foot, 8-by-12 white ash beam is not an easy lift by hand.). In addition, we consider the tractor a life-long investment. It has become my "on-site crew" in so many ways.

I wouldn't say that building a cordwood masonry earth-sheltered home with a living roof is for everyone. For the independent, alternatively minded builder with plenty of gumption, however, it is an extremely rewarding way to go.

Rob has succinctly outlined the advantages of cordwood masonry in *Cordwood Building: The State of the Art*, timber framing in *Timber Framing for the Rest of Us*, and living roofs in this current work. For us, one of the primary advantages of using these simple building systems is that virtually anyone is able to learn the techniques with just a few hours of instruction, and, with attention to detail, to do it well. In our case, we now have a home and a garage built with the participation of family and friends, enduring testaments to wonderful relationships.

Fig. 12.5: Mark and Mary's 40 × 40' home in Michigan has a living roof and is earth-sheltered on the north side, and on part of the east and west sides.

We think of our home as a manifestation of a way of looking at life. After having ushered it from concept to completion in the past few years it has become something more than a shelter; it feels like sanctuary. We are proud to have overcome obstacles that can get in the way of anyone who doesn't do the standard thing in the standard way. We've met with plenty of skepticism and, in some cases, outright derision from the imaginatively challenged. Yet, without exception, everyone who has walked through our hand-made cedar door has paused to marvel at what we have accomplished.

Perhaps, the mechanical inspector said it best in the wake of his final inspection:

"This is the most beautiful home in Emmet County, and I've been inside them all."

Appendix A: Radon

(The following article is adapted from one of my previous works. The author is indebted to Ned Doyle of Etowah, North Carolina for his careful review of the article and who made several valuable comments and suggestions. Ned is recognized by the US Environmental Protection Agency as a radon measurement and mitigation specialist and currently reviews radon training programs for the National Environmental Health Association.)

Radon is colorless odorless radioactive gas found in certain geological strata, including, but – not limited to – granite, shale, and gravelly subsoils. Its greatest health threat is increased risk of lung cancer. Breathing 20 picocuries per liter is considered to increase the risk of lung cancer about as much as smoking two packs of cigarettes a day. Homes can be built to greatly decrease the amount of radon entering the structure. In fact, in many parts of the country, attending to radon is a code requirement, and a careful reading of the appropriate sections is advised. Radon can also be greatly reduced in a home after it is built, albeit with difficulty. Listen:

Before radon gas was a well-known issue, my friends Pete and Eileen Allen built a partially earth-sheltered home with a very tight super-insulated geodesic dome above grade. They built the home over the well so that they would not have to worry about frozen pipes during our severe North Country winters. A woodstove provided in-floor heating to their concrete slab.

A couple of years after moving in, Pete's brother, a physician, was concerned about the possibility of radon in the home and gave my friends a radon-testing device. This may have been the best present they ever received. Test results showed that the winter level of radon in the home was about 40 picocuries per liter (pCi/l), ten times the "action level" recommended by the US Environmental Protection Agency. At 4 pCi/l, the EPA recommends that remedial work be done to lower radon levels.

Pete and Eileen, organic gardeners and clean-air proponents, didn't like the idea of their home having 30 times more risk of lung cancer than a non-smoker living in a low-radon home, according to research.

About the same time that the Allens were battling with their radon problem, Jaki and I were thinking of moving our oldest son to a bedroom in the fully earth-sheltered part of Earthwood, on the first floor. We were particularly concerned because we'd included into the walls of the home a number of "earth tubes" – sometimes called "cool tubes" – as an aid to summertime cooling. (Note: We now feel that the earth tubes were unnecessary and potentially problematic, and so do not include them as construction details in this book.) We wanted to be sure that we would not be subjecting a 9-year-old to the equivalent of smoking a pack of cigarettes every night while he slept. We sent for a test kit, and were relieved when the results came back at just 0.4 pCi/l, a tenth of the EPA's action-level figure.

But what about our friends? They tested again in the summer and found that radon levels were between 1 and 2 pCi/l, well within safety limits. But, the following winter, levels were high again. Remember that their dug well was right in the house with them and that they used wood heat. Pete noticed that cobwebs near the top of the well were rising up substantially whenever the woodstove was burning. Given the seasonal nature of their radon problem, our friends surmised that the woodstove was pulling radon gas up from the well.

The uppermost ten feet of the well was a 30-inch-diameter metal culvert. Pete sealed the well by pouring concrete in the bottom of the culvert. And he carefully sealed up all other potential sources of soil gasses, such as the house-drain cleanout. Finally, to decrease the negative pressure while the stove was in use, he opened a direct-air inlet to the exterior of the building. The efforts paid off, and the home's winter radon levels fell to the EPA's action level figure, or just below. The couple figures that the average level for the year is about 2 pCi/l. All existing homes – airtight or drafty – should be tested for radon, according to Ned Doyle.

Before you build a new earth-sheltered home, find out if the building site holds a lot of radon gas. Generalized maps of "radon areas" are only somewhat helpful, however. One building lot may have high levels of radon gas while a second lot just 100 yards away may have no radon at all. The only way to be sure is to conduct your own soil test. One company that sells a site-testing kit is Airchek, listed in Appendix B. Set up the little cardboard tent cubicle they supply on the subsoil itself. If you are planning a deep-hole test on site for septic design (or other) purposes, this would be a good time to check for radon. Send in the charcoal packet to Airchek and they will provide an analysis. Ned Doyle adds, "Remember that even low soil concentrations can end up as high radon levels in a finished home, depending on the pathways and pressure differentials between the home and the ground."

How serious a problem is radon gas? The EPA estimates that eight million American homes may contain unsafe radon levels. The National Cancer Institute considers exposure to radon to be the single greatest cause of lung cancer after smoking, accounting for an estimated 20–30,000 deaths a year in the US. Because an earth-sheltered home has a larger surface area in contact with the soil than a house on the surface, we should be particularly careful. The president of Airchek has said that, "We've yet to see an underground house that doesn't have some radon in it." Ned, who used to work for Airchek, adds that "all homes have at least background levels of radon." Finally, your state building code may require safeguards against radon contamination. So, what to do?

Again, I consulted my friend Ned Doyle.

"The best course of action," says Ned, "is to incorporate radon ameliorating techniques from the very beginning. It's easier and cheaper to avoid the problem than to fix it later. The US Environmental Protection Agency has free information on radon-resistant strategies for all types of housing. The goal is to facilitate the outgassing of any radon that finds its way into your home. An added benefit to these approaches is that even if you have no radon, moisture, humidity, mold, and other indoor air quality issues are often positively addressed. Many homes with radon sub-slab venting systems have been able to stop or greatly reduce their use of dehumidifiers and improve indoor air quality significantly.

Your county or state health departments are good sources of information. They may be able to tell you something about the incidence of radon in your area, as well as the names of reputable testing agencies. And the EPA has updated its free pamphlet on radon called "A Citizen's Guide to Radon." See the Annotated Bibliography for contact information.

What if testing indicates that your building site is "hot" with radon? There are building methods and products designed to keep radon gasses out of the home by venting the immediate surrounding soils directly into the atmosphere. Colbond Corporation (the manufacturers of Enkadrain® discussed in Chapter 9) makes a product called Enkavent®, essentially a half-inch crush-proof mesh which surrounds the building's fabric and carries the offending gas to the exterior by way of a vent stack, without the need for a fan. They will provide detailing with their product, which, obviously, must be one of the first systems installed during new home construction.

Ned Doyle suggests: "At least 4 inches of crushed stone under any slab makes sub-slab ventilation almost guaranteed. Homes with this underslab layer, where radon is later discovered to be a problem, can usually have the problem taken care of for $1,000 or less."

In addition, you will want to build with a view to eliminating all possible cracks in the fabric. A Bituthene®-like membrane will seal against infiltration through walls, while a layer of 10-mil polythene under the entire footings and floor area will greatly diminish the likelihood of gasses from below the floor. Avoid the use of expansion joints between the footings and floor, or between sections of the slab, as these can be sources of infiltration. The common "40-foot" rule for the requirement of expansion joints means that the 40-by-40 Log End Cave plan is the largest home which can be built without expansion joints. If you absolutely must include expansion jointing, tool and caulk the joint thoroughly. And, if there is significant radon in the soil, be sure to include a venting system such as Enkavent® or equal. Ned suggests the 4-inch crushed stone layer can work great, too, and could be less expensive.

If you are planning on a woodstove as heat, be sure to provide a direct source of outside combustion air. This is good practice, whether or not there is radon present. If radon is a concern, use at least a 6-inch-diameter solid-walled pipe, well sealed at any joints.

In conclusion, know your enemy. If in doubt as to whether or not you are in a high-risk area, conduct a site test. The presence of granite, shale or phosphate near the surface can all be indicators of radon, but the gas can occur in other strata as well. Ned's advice is to incorporate radon-resistant techniques in any case. We can all agree that if site testing reveals radon, a radon venting system must be designed into the home, as it is much more expensive to retrofit an earth-sheltered home (or basement) after it is built. Build with informed confidence. Finally, do not allow smoking in the home when even so-called "safe" levels of radon gas is present, because radon actually grabs onto smoke particles, greatly exacerbating the associated health risks.

APPENDIX B: RESOURCES

Author's note: The following contacts were checked for accuracy in October of 2005. However, companies sometimes move, are sold, or go out of business. If the information herein draws a blank, use search engines on the Internet for the latest contact information. Telephone area codes frequently change. I have not listed e-mail addresses because these, too, change frequently. Contact e-mails can almost always be found at the websites listed.

EDUCATION

British Earth Sheltering Association
   C/o David Woods (Secretary),
   4 Station Road, Coelbren,
   Neath, Wales, SA10 9PL
   Telephone: 011 44 1639 701 481 or
   C/o Peter Carpenter,
   Caer Llan Berm House, Lydart,
   Monmouthshire, Wales, NP25 4JS
   Telephone: 011 44 1600 860 359
   Website: www.besa-uk.org

The author has had many a delightful conversation with David and Peter about earth sheltering, stone circles, and other esoteric topics, usually over a pint of real ale. B.E.S.A. has continued to promote earth sheltering in the U.K., while similar American associations have fallen by the wayside. Lots of information on their website.

Earthwood Building School
   366 Murtagh Hill Road,
   West Chazy, NY 12992
   Telephone: (518) 493-7744
   Website: www.cordwoodmasonry.com

Conducts workshops in cordwood masonry, timber framing, and earth-sheltered housing, in northern New York, and around the world. Cordwood masonry is an excellent low-cost choice for above-grade portions of earth-sheltered homes. Rob does individual design consultations at $60/hour (2006.)

Malcolm Wells

    673 Satucket,

    Brewster, MA 02631

    Telephone (book sales only): (508) 896-6850

    Website: www.malcolmwells.com

Malcolm Wells under Education? Sure. Mac has done more to educate the world about underground housing than any other single individual. As if his entertaining books and articles were not enough, he now has this wonderful website. Spend a few hours cruising around the various contacts on his "Resources" page. Mac no longer consults or practices architecture, but he still sells his books, and that's a good thing.

Mole Publishing Co.

    Route 4, Box 618,

    Bonners Ferry, ID 83805

    Telephone: (208) 267-7349

    Website: www.undergroundhousing.com

Long-time co-conspirator Mike Oehler outflanks me on low-cost underground housing. Yes, he's even cheaper than I am. His *$50 and Up Underground House Book* is one of the best selling books in the field since 1978, and is still going strong. Check his site, and tell him Rob sent you.

FASTENERS

Cleveland Steel Specialty Company

    26001 Richmond Road,

    Bedford Heights, OH 44146.

    Telephone: (216) 464-9400 or (800) 251-8351

    Website: www.ClevelandSteel.com

Makers of a variety of wood-to-wood and wood-to-concrete connectors.

GRK Canada, Ltd.

    1499 Rosslyn Road,

    Thunder Bay, ON P7E 6W1 Canada.

    Telephone: (800) 263-0463

    Website: www.grkfasteners.com

Distributors of GRK Fasteners, including extremely strong and tough long screws for joining heavy timbers. Their smaller screws are excellent, too.

Olympic Manufacturing Group, Inc.

    153 Bowles Road,

    Agawam, MA 01001.

    Telephone: 1 (800) 633-3800

    Website: www.olyfast.com

Manufacturers of TimberLok™ screws for joining heavy timbers.

Simpson Strong-Tie Company, Inc.
    4120 Dublin Boulevard, Suite 400,
    Dublin, CA 94568.
    Telephone: (925) 560-9000 or (800) 999-5099
    Website: www.strongtie.com

Makers of a full line of wood construction connectors and fasteners for tying wooden members down to foundations.

USP Lumber Connectors
    703 Rogers Drive,
    Montgomery, MN, 56069-1324
    Telephone: (800) 328-5934.
    Website: www.USPconnectors.com

Manufactures a full line of lumber connectors.

INSULATION

Dow Chemical Company
    Styrofoam Brand Products,
    2020 Willard H. Dow Center,
    Midland, MI 48674
    Telephone: (989) 636-1000
    Website: www.dow.com/styrofoam/index.htm

Manufacturers of Styrofoam® brand extruded polystyrene insulation.

Styro Industries, Inc.
    P.O. Box 8098,
    Madison, WI 53708
    Telephone (Customer Service Department):
    (888) 702-9920
    Website: www.styro.net

Manufacturers of Flexcoat, a pre-mixed brush-on-grade protective coating for rigid foam insulation, and Tuff II, a trowel-grade product for the same purpose. Styro also makes FP Ultra Lite rigid foam foundation insulation panels, with either an aggregate (pebbledash) or stucco finish. It is important to protect all exposed rigid foam from UV and other deterioration. Styro products are a proven option. *ACE Hardware*, *Lowes*, and *True Value* are just three of many national distributors listed on their site.

T. Clear Corporation
    3255 Symmes Road,
    Hamilton, OH 45015
    Telephone: (800) 833-6444
    Website: www.tclear.com

Manufactures Lightguard® and Heavyguard® ballasted Styrofoam® panels as a waterproofing membrane protection system. Also makes Thermadry® below-grade drainage panels, which combine insulation and drainage in the same application.

LIGHT WELLS AND SKYLIGHTS

Solatube International, Inc.
2210 Oak Ridge Way,
Vista, CA 92081
Telephone: (800) 966-7652
Website: www.solatube.com

Manufacturers Solatube® light wells for bringing plenty of natural light into otherwise dark spaces.

Velux America Inc.
450 Old Brickyard Road,
Greenwood, SC 29648-5001
Website: http://www.veluxusa.com

Manufactures Velux skylights and Sun Tunnel light wells.

SURFACE-BONDING CEMENT

Bonsal American
8201 Arrowridge Boulevard,
Charlotte, NC 28273
Telephone: (704) 525-1621 or (800) 738-1621
Website: www.bonsal.com

Makers of Sakrete® Surface Bonding Cement and other cement-based products.

Conproco
17 Production Drive,
Dover, NH 03820
Telephone: (800) 258-3500
Website: www.conproco.com

Makes Structural Skin and Foundation Coat surface bonding cement and other related products.

Degussa Building Systems
889 Valley Park Drive,
Shakopee, MN 55379
Telephone (corporate office): (952) 496-6000 and (technical support, customer service): (800) 433-9517
Website: www.chemrex.com

Makes Thoro® Surface Bonding Mortar and Acryl-60 bonding agent, a very handy product for use with surface bonding and other masonry applications. I use it for bonding roofing slate to freshly poured concrete. Also supplies water-sealing cement products, such as Thoroseal®, and cement-patching material (Thorite® 400.)

Quikrete
One Securities Centre,
3490 Piedmont Road, Suite 1300,
Atlanta, GA 30305
Telephone: (404) 634-9100
Website: www.quikrete.com

Makes Quikwall® surface bonding cement and dozens of other cement-based products.

WATERPROOFING MEMBRANES AND DRAINAGE MATERIALS

American Hydrotech, Inc.
    303 E. Ohio Street,
    Chicago, IL 60611.
    Telephone: (800) 877-6125
    Website: http://www.hydrotechusa.com

Their Monolithic Membrane 6125® (MM6125) is a thick liquid-applied "self-healing" membrane made of refined asphalts and synthetic rubbers. It must be professionally applied. They also make the Hydrodrain® line of drainage composites and Thermaflo®, which combines drainage with Dow Styrofoam® insulation in one product.

American Wick Drain Corporation
    1209 Airport Road,
    Monroe, NC 28110
    Telephone: (800) 242-9425 or (704) 238-9200
    Website: www.americanwick.com

Makers of Amerdrain® composite drainage material, at a good price.

Carlisle Coatings & Waterproofing
    900 Hensley Lane,
    Wylie, TX 75098
    Telephone: (800) 527-7092
    Website: www.carlisle-ccw.com

Manufactures MiraDrain drainage composite and a variety of waterproofing membranes, including EPDM, butyl rubber, MiraDry sheet membranes, and MiraClay, based on bentonite. On their website, you can punch in your state, county, and product of interest and the site will return one or more nearby representatives.

Cetco
    1500 West Shure Drive,
    Arlington Heights, IL 60004
    Telephone: (847) 392-5800 or (800) 527-9948
    Website: www.cetco.com

Manufactures Volclay waterproofing panels, made with bentonite clay; Aquadrain composite drainage matting; and LDC 60, a liquid-applied modified polyurethane waterproofing membrane. In October of 2005, navigating the *Cetco* website took some perseverance.

Colbond, Inc.
    P.O. Box 1057, Sand Hill Road,
    Enka, NC 28728
    Telephone: (800) 365-7391 or (828) 665-5050
    www.colbond-usa.com

Makes the Enkadrain® drainage matting used at Earthwood's Stoneview guesthouse, Enkaroof® VM designed for pre-vegetated green roof applications, Enkavent® radon control matting, SubSeal-40™ and SubSeal-60™ waterproofing membranes, and Enkadri & Drain BTM, a one-step waterproofing and drainage system.

Grace Construction Products

    Division of W.R Grace and Co., Inc.,

    62 Whittemore Avenue,

    North Cambridge, MA 02140

    Telephone: (617) 876-1400.

    Website: www.graceconstruction.com

Address Note: This is a huge company. If you have Internet access, the best way to get information is to go to their website below, identify the type of product in which you are interested (e.g., Structural Waterproofing), and your state. The name and phone number and e-mail address of a representative will come up. Manufacturers of the various Bituthene® waterproofing membranes and drainage composites.

Soprema, Inc.

    310 Quadral Drive,

    Wadsworth, OH 44281

    Telephone: (330) 334-0066

    Website: www.soprema.com

Soprema Canada

    1640 Rue Haggerty,

    Drummondville, PQ J2C 5P8 Canada

    Telephone: (819) 478-8163 or (800) 567-1492

Manufactures a variety of waterproofing membranes for use with green roofs and below-grade applications. Also makes Sopradrain drainage composite. Chris Dancey (Chapter 8), used a Soprema membrane on her green roof.

Tremco Incorporated

    3735 Green Road,

    Beachwood, OH 44122

    Telephone: (800) 321-7906

    Website: www.tremcosealants.com

Suppliers of a full line of waterproofing membranes, including a trowel-grade bentonite product called Paramastic. Also makes drainage products.

MISCELLANEOUS

Airchek

    1933 Butler Bridge Road,

    Fletcher, NC 28732-9365

    Telephone: (828) 684-0893 or (800) AIR-CHEK

    Website: www.radon.com

Leading supplier of radon gas testing kits and equipment.

Amvic, Inc.

    501 McNicoll Avenue,

    Toronto, ON M2H 2E2 Canada

    Telephone: (877) 470-9991

    Website: www.amvicsystem.com

Makes Amvic Insulating Concrete Forms. Mark Powers built his 40-by-40-foot Log End Cave type of home (Chapter 12) using this poured wall system. He and Mary are pleased with the results.

Emory Knoll Farms, Inc.
    3410 Ady Road,
    Street, MD 21154
    Telephone: (410) 452-5880
    Website: www.greenroofplants.com

Supplies sedum and other plants suitable for lightweight living roofs. A family-owned business, Ed Snodgrass represents the sixth generation farming the same land. The website has good links to other Green Roof sites.

Formworks Building, Inc.
    P.O. Box 1509,
    Durango, CO 81302
    Telephone: (970) 247-2100
    Website: www.formworksbuilding.com

This is the company that constructed the underground vaults at Jim Milstein's home in the color section. Dale Pearcey, who started *Formworks*, has designed and built quite a number of earth sheltered homes, using the sprayed on concrete over vaulted forms. Many can be seen at the website.

Greenroofs-dot-com
    (No physical address or phone number)
    Website: www.greenroofs.com

This is an excellent green roof site. Their directory will put you in touch with other suppliers of useful products, too numerous to list here.

Selkirk, LLC
    P. O. Box 831950,
    Richardson, TX 75083-1950
    Telephone: (800) 992-8368 or (972) 943-6100
    Website: www.selkirkusa.com

Selkirk's Metalbestos chimney system for woodstoves is safe, easy to install, and compatible with living roofs.

Underground Gardens
    5021 West Shaw Avenue,
    Fresno, CA 93722
    Telephone: (559) 271-0734
    Website: www.undergroundgardens.com

Baldasare Forestiere built his hand-carved home, now called the *Underground Gardens*, during the first half of the 20th century. Open to the public. See Chapter 1.

The Wiremold Company
    60 Woodlawn Street,
    West Hartford, CT 06110
    Telephone: (860) 233-6251
    Website: www.wiremold.com

Wiremold Canada
    850 Gartshore Street,
    Fergus, ON N1M 2W8 Canada
    Telephone: (519) 843-4332

The use of *Wiremold* and other surface-mounted electrical conduits enables the builder to build first and wire later. Circuits can be repaired, added to, and subtracted from, without difficulty.

Appendix C: Stress Load Calculations

Span tables will serve for roof design with most structures. With heavy roofs, such as earth roofs, adequate tables are very hard to find. This Appendix shows how to check the girders and rafters in a heavy roof design for shear and bending. Once you have followed through the example, and understand where all the numbers have come from, you should be able to use the formulas and procedures to check other rectilinear designs. Using this Appendix requires familiarity with basic algebra, specifically the ability to substitute numbers for letters in a formula, and to solve for a single unknown. It is important to keep track of the units (feet, pounds, etc.) as you solve the equations. Problem: Test the 40- by 40-foot Log End Cave Plan for the shear and bending strength of the rafters and girders as designed. (The posts and planks are the strong — in some ways overbuilt — components of this design.) A portion of the plan, enough for our purposes, is shown in Fig. C.1. Girders are labeled "beams" on the plan. The plan is based upon simple 10-foot-square sections, repeated sixteen times, like a chessboard with just four squares on a side. Only six complete sections are shown in the portion reproduced here. Here are the givens:

Design Load:

- Earth roof, saturated; 8 inches at 10 pounds/inch/SF . . . . .80 pounds/SF
- Crushed stone drainage layer; 2 inches at 10 pounds/inch/SF . . . . .20 pounds/SF
- Snow load by code, Plattsburgh, NY . . . . .70 pounds/SF
- Structural load, typical for scale of heavy timber structure (includes timbers, planking, membrane, and insulation) . . . . .15 pounds/SF
- Total maximum load . . . . .185 pounds/SF

Kind and Grade of Wood:

Different species of woods have different stress load ratings, and the lumber grade has a large impact on the ratings, too, as can be seen from these few examples from Architectural Graphic Standards:

Type of Wood	Grade	f_b[1]	f_v[2]
Douglas Fir, Inland Region	Select Structural	2,150	145
Douglas Fir, Inland Region	Common Structural	1,450	95
Eastern Hemlock	Select Structural	1,300	85
Eastern Hemlock	Common Structural	1,100	60
Southern Pine	#1 Dense	1,700	150
Southern Pine	#2	1,100	85

[1] unit stress for bending in pounds per square inch
[2] unit stress for shear in pounds per square inch

For our example, all timbers are assumed to be Douglas Fir (Inland Region, Common Structural) with the following stress load values:

- Unit stress for bending (f_b) of 1,450 pounds per square inch
- Unit stress for horizontal shear (f_v) of 95 pounds per square inch

These are moderate values, incidentally, similar to Eastern spruce and red and white pine. See *Architectural Graphic Standards, The Encyclopedia of Wood, A Timber Framer's Workshop* and other engineering manuals for stress load ratings for a variety of woods.

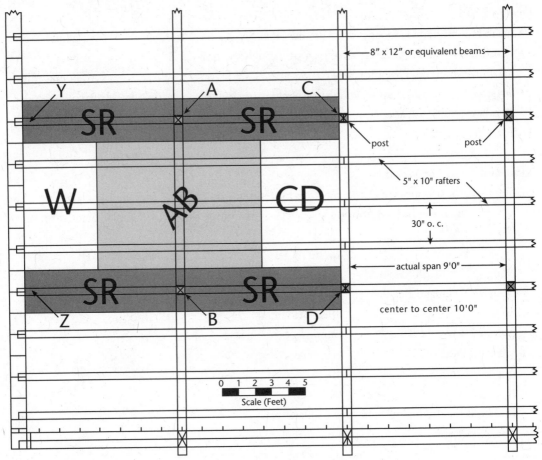

Fig. C. 1:
A portion of the
40 × 40' Log
End Cave plan.

Cross-sectional Dimensions (b and d):

- Rafters are "five-by-tens," that is, they are five inches (12.7 cm) in breadth (b) and ten inches (25.4 cm) in depth (d).
- Girders ("beams" on the plan) are "eight-by-twelves," that is, they are eight inches (20.3 cm) in breadth (b) and twelve inches (30.5 cm) in depth (d).

Frequency (Spacing):

- Rafters are 30 inches o.c., that is: 30 inches (76 cm) is the center-to-center spacing for adjacent members.

- Girders are 10 feet o.c., that is: ten feet (3 m) is the center-to-center spacing between parallel girders or between girders and the side walls.

Span (L):

- Spans are nominally ten feet (3 m) for both girders and rafters. Actual clear spans, from the edge of one support to the edge of another, is closer to nine feet (2.75 m), but 10 feet is the number used in place of L (span) in the example.

Nomenclature:

- "Beam" refers to both rafters and girders
- "Simple Span" means that a beam is supported only at its ends.
- "Double Span" means that a beam is supported at its ends, and also at its midpoint.

A = Cross-sectional area (b times d) of beam in square inches

b = Breadth of beam, in inches

d = Depth of beam, in inches

f_b = Allowable unit stress for bending in pounds per square inch

f_v = Allowable unit stress for shear in pounds per square inch

L = Length of span in feet

M = Bending moment in foot-pounds or inch-pounds

M_x = Bending moment at the two midspans on a double-span beam

PSF = pounds per square foot

R = Reaction at supports

S = Section modulus of cross-section of beam in inches cubed

V = Total shear allowable or actual

w = Load or weight per linear foot on beam, in pounds

W = Total uniform load or weight on beam, in pounds

Algebraic Operations:

$/$ = The division sign. The value before the division sign is divided by the value after it.

$6(8) = 48$ or $(6)(8) = 48$ means "6 times 8 equals 48." The "times" sign is implied.

$bd = A$ means "b times d equals A." Again, multiplication is implied.

Formulas used with Simple-Spans:

$R = V = wL/2 \qquad M = wL/8 \qquad S = bd^2/6$

$S = M/f_b \qquad f_v = 3V/2A \qquad V = wL/2$

Formulas used with Double Spans:

$R_1 = V_1 = R_3 = V_3 = 3wL/8 \qquad R2 = 2V_2 = 10wL/8$

$V_2 = 5wL/8 \qquad M_x = 9wL^2/128$

We have now listed the five variables for structural design for shear and bending, and we have all the nomenclature and formulas that we need. (See also Chapter 2 of *Timber Framing for the Rest of Us*).

Now we want to find out if the structure as designed – particularly the rafters and girders – is of adequate strength for both shear and bending to support the design load of 185 pounds per square foot (903 kilos per square meter).

1. Calculating roof load for bending for rafters, simple-span.

(That is, all rafters are about ten feet long, and join over girders.)

$S = bd^2/6 = (5")(10")^2/6 = 83.3$ in^3 (Section modulus is measured in "inches cubed")

$f_b = 1,450$ psi (pounds/square inch), given above for Douglas Fir, Inland Region, Common Structural

$S = M/f_b$. By transposition: $M = S(f_b) = 83.3$in^3 (1450 lb/in^2) $= 120,785$ in. lbs

This is the bending moment in "inch-pounds". To derive the more convenient "foot-pounds," we need to divide by 12 in/ft, because there are 12 inches in a foot.

So:

$120,785$ in. lbs divided by 12 in/ft $= 10,065$ foot-pounds

$L = 10'$ (given). $\qquad M = wL^2/8$. $\qquad$ By transposition: $w = 8M/L^2$

Substituting for M and L: $w = 8(10,065$ ft lbs$)/100$ ft$^2 = 805$ lbs/ft

That is: 805 pounds per linear foot. We haven't got pounds per square feet quite yet. If rafters were on 12-inch centers, they could support 805 pounds per square foot (3,930 kilos per square meter). A linear foot would translate to a square foot in this special case. But our example calls for rafters on 30-inch centers, so we need to make the following adjustment:

12"/sq. ft. divided by 30" = 0.4 ft.(805 lbs/ft) = **322 PSF** allowable

Think of it this way: There are only 40 percent (0.4) as many rafters on 30-inch centers as on 12-inch centers. As the impact of frequency is a direct proportional relationship to strength, the rafters on 30-inch centers will support only 40 percent of the load, everything else remaining the same.

The specified rafters, on simple span, will easily support the 185 PSF required.

Now let's try it on a double span. We'll use 20-foot-long rafters, supported at each end, but also at the middle by a girder.

2. **Calculating roof load for bending for rafters, double-span. (That is, all rafters are about twenty feet long, and supported at midspan by a girder.)**

Maximum allowable bending moment (M) = 10,065 foot pounds, from calculation (1) above.

M_x = Bending moment at the two midspans on a double-span beam

$M_x = 9wL^2/128$ (from formulas above)

$W = 128 \, M_x /9L^2$ $w = 128(10,065 \text{ ft. lbs.})/9(10 \text{ ft})^2 = 1431$ lbs./ft

Again, this is "pounds per linear foot." We make the same adjustment that we made at the end of calculation (1) above:

12"/sq. ft divided by 30" = 0.4 ft.(1,431 lbs/ft) = **572 PSF** allowable

Using a single 20-footer, supported halfway, increases bending strength by quite a bit, but this value is far stronger than it needs to be. Now, let's test rafters for shear.

3. **Calculating roof load for shear on simple-span.**

f_v = 95 psi (pounds/square inch), given above for the same grade of Douglas Fir

A = bd = 5"(10") = 50 inches squared (In this case, the same as "square inches.")

$f_v = 3V/2A$. By transposition: $V = 2Af_v /3 = 2(50 \text{ in}^2)(95 \text{ lbs})/3(\text{in}^2) = 3,167$ pounds maximum allowable (V is "total shear allowable")

V = wL/2. So, w = 2V/L = 2(3,167 lbs)/10 feet = 633 pounds per (linear) foot

But, again, rafters are not on 12" centers, but are actually 30 inches o.c. Making the adjustment: 12"/30" = 0.4 0.4(633) = **253 PSF** allowable, another good strong number.

4. Calculating roof load for shear on a double-span rafter.

Maximum allowable shear (V) = 3,167 pounds from calculation (3) above.

Shear at ends (R_1 and R_3): V = 3wL/8. Transposed: w = 8V/3L

W = 8(3,167 pounds)/3(10 feet) = 845 pounds per lineal foot

Rafters are 30 inches o.c., so: 12"/30" = 0.4; 0.4(845) = 338 pounds per square foot

That is the shear strength at the ends, at R_1 and R_3. But, at R_2, the center support, the situation is a little different:

Shear at middle (R_2): V = 5wL/8. Transposed: w = 8V/5L

W = 8(3,167 pounds)/5(10 feet) = 507 pounds per lineal foot

Rafters are 30 inches o.c., so: 12"/30" = 0.4; 0.4(507) = **203 PSF**, still more than the 185 PSF foot required.

Now let's do the girders, and we'll just do them for single-span because 20-foot-plus eight-by-twelve girders are really a bit extreme. Plus, as we know, they will not only be easier to install as two 10-footers, but the shorter pieces will actually be stronger on shear.

5. Calculating roof load for bending on the single-span 8 × 12" Douglas fir girders of this example. The load from the rafters is symmetrically placed along the girder at regular 30-inch spacings, so it is reasonable to use the same formulas we used for single-span rafters.

S = $bd^2/6$ = (8")(12")2/6 = 192 in^3 (Section modulus is measured in "inches cubed")

f_b = 1450 psi (pounds/square inch), given above for Douglas Fir, Inland Region, Common Structural

S = M/f_b. By transposition: M = S(f_b) M = 192in^3(1,450 lb/in^2) = 278,400 in. lbs

This is the bending moment in "inch pounds". To derive the more convenient "foot pounds," we need to divide by 12 in/ft, because there are 12 inches in a foot.

So:

278,400 in. lbs divided by 12 in/ft = 23,200 foot pounds

L = 10' (given) M = $wL^2/8$. By transposition: w = $8M/L^2$

Substituting for M and L: w = 8(23,200 ft lbs)/100 ft^2 = 1,856 pounds per linear foot

The girder can support 1,856 pounds per linear foot, or 18,560 pounds in all, if the load is fairly constant along its length. But for what portion of the roof is the girder responsible? Look again at Fig. C.1 on page 236. The area AB is the area for which girder A-B is responsible. Area CD is part of the area carried by the girder C-D. Area W is carried by the block wall. The two areas labeled SR are carried by the special rafters labeled Y and Z. Y and Z are special because their loads are carried directly down through the girders to the posts, adding no bending stresses to the girder. The area AB is 10 feet by 7.5 feet or 75 square feet. So, the total allowable carrying capacity of the girder (18,560 pounds in all) divided by the square footage for which it is responsible (75 square feet) results in the allowable load per square foot, assuming an equally distributed load. 18,560 pounds/75 SF = **247.5 PSF**. Still a good number, as it is higher than 185 PSF. Now, what about girders on shear?

6. **Calculating roof load for shear on the single-span 8 × 12" Douglas fir girders of this example. The load from the rafters is symmetrically placed along the girder at regular 30-inch spacings, so it is reasonable to use the same formulas we used for single-span rafters.**

f_v = 95 psi (pounds/square inch), given above for the same grade of Douglas Fir

A = bd = 6"(12") = 96 inches squared (In this case, the same as "square inches.")

f_v = $3V/2A$. By transposition: V = $2Af_v/3$ = 2(96 in^2)(95 lbs)/3(in^2) = 6,080 pounds maximum allowable (V is "total shear allowable")

To get the shear strength at the ends of a single-span rafter, use:

V = $wL/2$ So, w = $2V/L$ = 2(6,080 lbs)/10 feet = 1,216 pounds per (linear) foot, or 12,160 pounds over 10 feet.

Again, the area for which girder A-B is responsible is area AB, or 75 SF.

12,160 pounds divided by 75 SF results in **162.1 PSF**, which is less than the desired carrying capacity of 185 PSF for the earth roof described. It doesn't look good. However, if we consider that the true girder clear span (between posts) is actually 9 foot 4 inches and substitute 9 foot 4 inches (9.333') for 10 feet in w = 2V/L, we get w = 1,303 pounds per linear foot, or 13,030 over 10 feet. Divided by 75 SF results in **173.7 PSF**, closer, but still a little short of the mark. What can we do?

Shear, unlike bending, is a direct linear relationship. The shortfall can be made up in variety of ways. These will all work:

A. Beef the girders up to 9 inches wide. Now A = 108 square inches instead of 96 square inches. This change increases the cross-sectional area of the girder – and its shear strength – by 12.5 percent because 12/96 = .125. Now, 173.7 PSF times 1.125 equals **195.4 PSF**, so we're good again.

B. Use a wood with a unit stress for horizontal shear at least 10 percent greater than the 95 psi for Douglas Fir (Inland Region, Common Structural). Any wood with an (f_v) of at least 105 psi would do nicely.

C. Shorten the girder clear span by 7 inches to 8 foot 9 inches (8.75'). This yields **185.3 PSF**, which is fine, as there are great safety factors built into these calculations. Just work to the numbers. You don't have to add an additional safety factor.

D. If you want to keep the plan as designed, you could always decrease the load by about 12 PSF, down to 173 PSF. Eliminate 1.2 inches of earth or crushed stone. Is this cheating? In point of fact, the stone and earth layers at Earthwood are really about 8.5 inches total, not 10 inches, so our load here is probably about 170 PSF. This is enough to maintain our living roof.

Incidentally, using a 20-foot girder, supported half way, weakens the plan unacceptably in terms of shear strength for the girders. While shear strength increases at the ends to 218 PSF, it decreases over the center support to 130 PSF. Strange, but true.

Disclaimer: The author is not an engineer. Use these exercises as a point of beginning, to get you into the ballpark. Always have your plan checked by a qualified structural engineer.

APPENDIX D: METRIC CONVERSION TABLE

Here are some handy conversions for measurements:

If you know:	× by:	To Find:	If you know:	× by:	To Find:
Inches	25.4	millimeters	Millimeters	0.039	inches
Inches	2.54	centimeters	Centimeters	0.39	inches
Feet	0.3048	meters	Meters	3.281	feet
Yards	0.9144	meters	Meters	1.094	yards
Miles	1.609	kilometers	Kilometers	0.622	miles
Fluid ounces	28.4	milliliters	Milliliters	0.035	fluid ounces
Gallons	4.546	liters	Liters	0.220	gallons
Ounces	28.350	grams	Grams	0.035	ounces
Pounds	0.4536	kilograms	Kilograms	2.205	pounds
Short tons	0.9072	metric tons	Metric tons	1.102	short tons
Acres	0.4057	hectares	Hectares	2.471	acres
inches^2	6.452	square centimeters	centimeters^2	0.155	square inches
yards^2	0.836	square meters	meters^2	1.196	square yards
yards^3	0.765	cubic meters	meters^3	1.307	cubic yards
miles^2	2.590	square kilometers	kilometers^2	0.386	square miles

Annotated Bibliography

Frost-Protected Foundations

Design Guide for Frost-Protected Shallow Foundations, prepared for the US Department of Housing and Urban Development (HUD) by NAHB Research Center, Upper Marlboro, MD, July 1994. Instrument number: DU100K00000 5897. 56 pages.

This document can be downloaded free at <www.huduser.org/Publications/PDF/FPSFguide.pdf>.

Labs, Kenneth, John Carmody and Jeffrey Chistian. *Builder's Foundation Handbook*, prepared for Oak Ridge National Laboratory, Oak Ridge, TN 37831. 124 pages.

Good detailing on radon, termites, and slab-on-grade. To follow all the recommendations in here would be very expensive and probably overkill. This document can be downloaded free at <www.ornl.gov/sci/roofs+walls/foundation/ORNLCON-295.pdf>.

The Editors of Fine Homebuilding. *Foundations and Concrete Work*. Taunton Press, 1998. ISBN: 1-56158-330-8. (Out of print as of January 2006).

Lots of good conventional and not-so-conventional stuff here. Good hard-to-find information on the rubble trench foundation and the slab-on-grade (floating slab) foundation. There are good articles on surface-bonded blocks and site layout. And the other 20-odd articles aren't bad either. Be sure to check the ISBN number when buying, borrowing, or ordering this book, as Taunton publishes a different book with the same title but different ISBN number.

Cordwood Building

Roy, Rob. *Cordwood Building: The State of the Art.* New Society Publishers, 2003. ISBN: 0-86571-475-4.

Up-to-date information about cordwood masonry by over 20 of the field's leading proponents. Cordwood masonry is the author's favorite low-cost building technique for above-grade walls in combination with earth sheltering. A good companion volume to the present work.

Stankevitz, Alan, Richard Flatau, Rob Roy and Dr. Kris Dick. *Cordwood and the Code: A Building Permit Guide*. Cordwood Construction, W4837 Schulz Spur Drive, Merrill, WI 54452. No ISBN number. Available also from Earthwood Building School. (See Appendix B).

Simply the best available document to help you smooth the way with your local building department for building with cordwood masonry.

TIMBER FRAMING

Roy, Rob. *Timber Framing for the Rest of Us: A Guide to Contemporary Post and Beam Construction*. New Society Publishers, 2004. ISBN: 0-86571-508-4.

This one is a detailed examination of using commonly available specialty screws and other mechanical fasteners to construct a strong and beautiful post-and-beam frame quickly and easily. The new sunroom at Earthwood serves as a step-by-step model to illustrate the techniques. Structural considerations are covered in detail, as well as wood procurement. This book works well in combination with the present work.

Chappell, Steve. *A Timber Framer's Workshop*. Fox Maple Press, 1999. Distributed by Chelsea Green. ISBN: 1-889269-00-x.

Describes the tools, structural considerations, design, roof framing, and the wood jointing details used in traditional timber framing. Chapter 11, "Building Math and Engineering," is enlightening.

CONVENTIONAL BUILDING

Nash, George. *Do-It-Yourself Housebuilding*. Sterling, 1995. ISBN: 0-8069-0424-0.

This huge well-written volume is the one book you need to answer questions concerning conventional house construction. It's all here: foundations, electric, plumbing, framing, and roofing.

EARTH-SHELTERED (UNDERGROUND) HOUSING

Dunnett, Nigel and Noel Kingsbury. *Planting Green Roofs and Living Walls*. Timber Press, 2004. ISBN: 0-88192-640-x.

A treatise from two English authors about the form and function of earth roofs around the world. Particularly strong on appropriate vegetation.

Hait, John N. *Passive Annual Heat Storage*. Rocky Mountain Research Center, Missoula, MT 1983. ISBN: 0-915207-99-1.

This out-of-print book is essential reading if you want an insulation umbrella surrounding the earth near your home, something Geoff Huggins did successfully on his earth-sheltered home near Winchester, Virginia.

Kern, Ken & Barbara, and Jane & Otis Mullan. *The Earth-Sheltered Owner-Built Home.* Mullein Press/Owner-Builder Publications, 1982. ISBN: 0-910225-00-1.

This out-of-print book has an excellent chapter on the hand-carved underground home and gardens of Baldasare Forestiere, mentioned in Chapter 1 of this book. There are detailed case-studies of three owner-built earth-sheltered buildings. You can also do a Google search on the Internet for info on Baldasare Forestiere.

Oehler, Mike. *The $50 and Up Underground House Book.* Mole Publishing Company, 1978, 1997. ISBN: 0-442-27311-8.

This underground classic describes the construction of basic but comfortable survival shelter at the lowest possible cost. Information not found in other books is presented in Mike's entertaining writing style.

Wells, Malcolm. *How to Build an Underground House.* Published by the author, 673 Satucket Road, Brewster, MA 02631. ISBN: 0-9621878-3-6.

Mac packs a lot of useful information in this hand-written 96-page volume. Takes the reader through the design and construction of a 1,200 square-foot earth-shelter that can be built on a flat or sloped lot.

Wells, Malcolm and Sam Glenn-Wells. *Underground Plans Book 1.* Published by Mac Wells, 673 Satucket Road, Brewster, MA 02631. No ISBN number.

This wide book – 42 inches fully extended! – has designs for nine different earth-sheltered houses and lots of good detailing. These plans are not intended to be construction documents, but are a great starting point for plans.

Wells, Malcolm. *The Earth-Sheltered House: An Architect's Sketchbook.* Chelsea Green, 1998. ISBN: 1-890132-19-5.

An entertaining idea book, still in print, and worth a look.

Also: Back in the 1980's, there was an active Underground Space Center at the University of Minnesota at Minneapolis. Unfortunately, this excellent department was disbanded in the 1990s. During their existence, the USC published three excellent and well-researched documents, and, although a lot of new products have come on the market in the past 20 years, the basic construction information and case studies are still useful today. Although all of their books are out of print, lots of copies were published by Van Nostrand Reinhold and so you may find copies at used book stores, libraries, or on the Internet. The three books are:

Earth-Sheltered Housing Design: Second Edition. 1985. ISBN: 0-442-28746-1.

The best and most thorough of the USC books, this one was long considered as the definitive work in the earth-sheltered housing field. Large, detailed, well-documented. Contains 17 case studies with basic plans. Lots of copies were printed, so it is not too difficult to find.

Earth Sheltered Residential Design Manual. 1982. ISBN: 0-442-28679-1.

Good overview of ESH. Strong on structural details, waterproofing, insulation, and regional considerations. Hard to find. Try an inter-library loan.

Earth Sheltered Homes: Plans and Designs. 1981. ISBN: 0-442-28676-7.

Contains 23 case studies from around the United States and Europe. Most of the studies have a short narrative, excellent color pictures, plans, and detailing drawings. Wonderful idea book, but rare. Again, try an inter-library loan.

MASONRY STOVES

Lyle, David. *The Book of Masonry Stoves: Rediscovering an Old Way of Warming.* Chelsea Green, 1997. ISBN: 1-890132-09-8.

A highly informative and entertaining book about heating with wood, especially with the varieties of masonry heaters used around the world. Not a how-to book, but still well worth a read.

Roy, Rob. *Building the Earthwood Masonry Stove.* Published by Rob and Jaki Roy, Earthwood, 366 Murtagh Hill Road, West Chazy, NY 12992. No ISBN number.

Contains 20 step-by-step construction images. Available as color slides or as 4-by-6-inch prints. A 3,000-word commentary is keyed to the images, explaining construction details.

RADON GAS

US Environmental Protection Agency. *A Citizen's Guide to Radon.* (EPA 402-K02-006, Revised May 2004).

Can be downloaded as a PDF file. Go to <www.epa.gov/radon/>, the EPA's main radon site (with lots of useful information in its own right) and then click on Publications. In addition to *A Citizen's Guide*, you can also download the newly revised *Home Buyer's and Seller's Guide to Radon* (EPA 402-K-05-005, May 2005), *Building Radon Out: A Step-by-Step Guide on How to Build Radon-Resistant Homes* (EPA 402-K-01-002, April 2001), and *Consumer's Guide to Radon Reduction* (EPA 402-K-03-002, Revised February 2003). For print versions, contact the US EPA/Office of Radiation and Indoor Air, Indoor Environments Division, 1200 Pennsylvania Avenue, NW, Mail Code 6609J, Washington, DC 20460. Or call (202) 343-9370.

SLIP-FORM STONE BUILDING

Stanley, Tomm. *Stone House: A Guide to Self-Building with Slipforms*. Stonefield Publishing, 2004. ISBN: 0473099705.

    While providing a wealth of information on related subjects, the author uses an instructional narrative to describe the process of building with stone and slipforms. This is the first book dedicated to slipform stone masonry in many years.

Schwenke, Karl and Sue. *Build Your Own Stone House Using the Easy Slipform Method*, Garden Way, 1975.

    Long out of print, this old classic can still be found on the Internet, and, of course, in libraries.

HUMANURE

Jenkins, Joseph. *The Humanure Handbook: A Guide to Composting Human Manure*. Chelsea Green, 2005. ISBN: 978-0-9644258-3-5.

    Joe's simple and ecological humanure composting system is to human septic disposal as earth sheltered housing is to housing: sensible. What we are doing with human waste, in land-destroying septic systems, is as bad as the human predilection towards paving the planet and roofing our homes with black petrochemicals.

INDEX

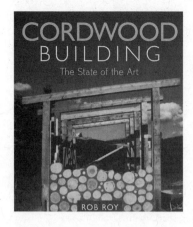

All you need to create beautiful buildings from sawmill left-overs — or even driftwood!

Cordwood Building: *The State of the Art*

Rob Roy

Cordwood masonry is an ancient building technique whereby walls are constructed from "log ends" laid transversely in the wall. It is easy, economical, esthetically striking, energy-efficient, and environmentally-sound. *Cordwood Building* collects the wisdom of over 25 of the world's best practitioners, detailing the long history of the method, and demonstrating how to build a cordwood home using the latest and most up-to-date techniques, with a special focus on building code issues.

A manual for all without traditional skills who want to build with timber framing

Timber Framing for the Rest of Us
A Guide to Contemporary Post and Beam Construction

Rob Roy

Timber Framing for the Rest of Us describes the timber framing methods used by most contractors, farmers, and owner-builders, methods that use modern metal fasteners, special screws, and common sense building principles to accomplish the same goal in much less time. And while there are many good books on traditional timber framing, this is the first to describe in depth these more common fastening methods. The book includes everything an owner-builder needs to know about building strong and beautiful structural frames from heavy timbers, including:

- the historical background of timber framing
- procuring timbers -- including different woods, and recycled materials
- crucial design and structural considerations
- foundations, roofs, and in-filling considerations
- the common fasteners.

A detailed case study of a timber frame project from start to finish completes this practical and comprehensive guide, along with a useful appendix of span tables and a bibliography.

About the Author

Rob Roy has been teaching alternative building methods to owner-builders since 1978, and has become well-known as a writer and educator in fields as disparate as cordwood masonry, earth-sheltered housing, saunas, stone circles and mortgage freedom. This is his thirteenth book for owner-builders. In addition, Rob has created four published videos, also for owner-builders. With his wife Jaki, Rob started Earthwood Building School in 1981. The couple teaches their building techniques all over the world, but their home is at Earthwood in West Chazy, New York. To learn more about Earthwood, visit them online at <www.cordwoodmasonry.com>.

If you have enjoyed *Earth-Sheltered Houses*
you might also enjoy other

BOOKS TO BUILD A NEW SOCIETY

Our books provide positive solutions for people who want to
make a difference. We specialize in:

**Environment and Justice • Conscientious Commerce
Sustainable Living • Ecological Design and Planning
Natural Building & Appropriate Technology • New Forestry
Educational and Parenting Resources • Nonviolence
Progressive Leadership • Resistance and Community**

New Society Publishers

ENVIRONMENTAL BENEFITS STATEMENT

New Society Publishers has chosen to produce this book on Enviro 100, recycled paper made with **100% post consumer waste**, processed chlorine free, and old growth free.

For every 5,000 books printed, New Society saves the following resources:[1]

48	Trees
4,348	Pounds of Solid Waste
4,784	Gallons of Water
6,240	Kilowatt Hours of Electricity
7,904	Pounds of Greenhouse Gases
34	Pounds of HAPs, VOCs, and AOX Combined
12	Cubic Yards of Landfill Space

[1]Environmental benefits are calculated based on research done by the Environmental Defense Fund and other members of the Paper Task Force who study the environmental impacts of the paper industry.

For more information on this environmental benefits statement, or to inquire about environmentally friendly papers, please contact New Leaf Paper – info@newleafpaper.com Tel: 888 • 989 • 5323.

For a full list of NSP's titles, please call **1-800-567-6772** *or check out our website at:*

www.newsociety.com

NEW SOCIETY PUBLISHERS